Frank Whittle

The Invention of the Jet

Andrew Nahum

This edition published in the UK in 2017 by
Icon Books Ltd, Omnibus Business Centre,
39–41 North Road, London N7 9DP
email: info@iconbooks.com
www.iconbooks.com

Originally published in 2004 and 2005 by Icon Books Ltd

Sold in the UK, Europe and Asia by
Faber & Faber Ltd, Bloomsbury House,
74–77 Great Russell Street,
London WC1B 3DA or their agents

Distributed in the UK, Europe and Asia by
Grantham Book Services, Trent Road,
Grantham NG31 7XQ

Distributed in the USA by
Publishers Group West,
1700 Fourth Street, Berkeley, CA 94710

Distributed in Canada by
Publishers Group Canada,
76 Stafford Street, Unit 300,
Toronto, Ontario M6J 2S1

Distributed in Australia and New Zealand by
Allen & Unwin Pty Ltd, PO Box 8500,
83 Alexander Street,
Crows Nest, NSW 2065

Distributed in South Africa by
Jonathan Ball, Office B4, The District,
41 Sir Lowry Road, Woodstock 7925

ISBN: 978-1-78578-241-1

Typesetting by Born Group

Printed and bound in the UK by Clays Ltd, St Ives plc

Contents

List of Figures and Plates	**iv**
Acknowledgements	**v**
About the Author	**vii**
Foreword	**viii**
Introduction	**1**
1 Whittle's Early Jet Ideas	**12**
2 The Formation of Power Jets Limited	**22**
3 Wartime Development	**50**
4 Even With All the Mistakes – Human and Otherwise	**57**
5 The 'Straight Through' Engine	**65**
6 The Origins of the Jet Engine Programme in the USA	**91**
7 The Nationalisation of Power Jets	**104**
8 A State-Owned Firm: Power Jets (R&D) Ltd	**115**
9 The End of Power Jets: Formation of the National Gas Turbine Establishment	**120**
10 Jet Modernity: the Comet	**127**
11 Comet Failure	**139**
12 The Leadership So Narrowly Missed	**148**
13 New Jets	**152**
Endnotes	
A Jet Without Whittle? Developments in the USA	157
The Jet in Germany	161
Jet Histories	165
Notes	**171**
Bibliography	**173**

List of Figures

1 The essential Whittle design 15
2 Diagrammatical view of an axial flow engine 20
3 A 'straight through' centrifugal compressor engine 67

List of Plates

1 Whittle in his study in 1946
2 The first ground test engine (the W.U.) at Power Jets' works in Lutterworth
3 The rotors of a Power Jets W.2/700 and a Metrovick F2 engine
4 Whittle and colleagues recreating the test of the first engine for the official film, *Jet Propulsion*
5 The Gloster-Whittle E.28/39
6 The headline story from the *Daily Express* for Friday 7 January 1944
7 The de Havilland Comet 1 prototype
8 Vickers VC10 airliner used for flight tests of the new Rolls-Royce RB 211

Acknowledgements

Over a number of years I have been very fortunate in being able to talk to a large number of people who were involved in wartime and post-war aeronautical research and development, some, sadly, no longer with us. A number of these were 'Reactionaries' – members of the informal association for members of Frank Whittle's team at Power Jets – while others have been from industry and from the government research establishments. Some are acknowledged directly in the text, although others are anonymous. However I thank them all warmly. Some years ago, the Reactionaries most kindly asked me to their 40th anniversary dinner, held at the wartime headquarters of Brownsover Hall, near Rugby, which started a number of these conversations.

At the Science Museum, my predecessor as Curator of Aeronautics, the late John Bagley, always encouraged my interest in engines and aeronautical history. Through many conversations he also gave an illuminating insight, from his own time as an aerodynamicist at the Royal Aircraft Establishment, Farnborough, into the culture of government defence research and into the relationship between industry and the research establishments.

It would have been impossible to undertake this study without the generous help of the Science Museum, and it is a pleasure to acknowledge the impetus given by Robert Bud, as Head of Research, and Lindsay Sharp as Director. Furthermore, Jon Tucker, Heather Mayfield and Tim Boon have given excellent encouragement.

Special thanks also to Simon Flynn, Jon Turney and Ruth Nelson at Icon Books whose comments and impulsion have helped tremendously.

Dedication

For Fiona

About the Author

Andrew Nahum is a curator, writer and historian. He is Keeper Emeritus at the Science Museum where he led the creation of the landmark gallery Making the Modern World (2000) and many other exhibitions including *Inside the Spitfire* (2005), *Dan Dare and the Birth of Hi-tech Britain* (2008) and *Churchill's Scientists* (2015). In 2017 he curated the exhibition *Ferrari – Under the Skin*, at the Design Museum, London.

He has written extensively on the history of technology and on design, and his books include *Alec Issigonis and the Mini* (Icon Books, 2005), a contribution to the history of British aeronautics in the Cold War, 'The Royal Aircraft Establishment from 1945 to Concorde' in *Cold War, Hot Science*, (Science Museum, 2002) and *Fifty Cars that Changed the World* (Design Museum, 2009 and 2016).

FOREWORD

Since the publication of earlier editions of this book, the remarkable phenomenon of the development of the jet engine under the pressures of the Second World War has benefited from renewed and welcome historical attention. However, the account here remains unique in that it sets specifically out to study the pervasive and sometimes misleading 'invention story' of the British jet engine, and to calibrate it against the available contemporary accounts and recollections from many of the actors involved in the drama – for drama it certainly was.

It is often said that Britain is good at inventing but bad at supporting its inventors. This book shows that wartime Britain was the reverse – a highly technocratic state eager to find scientific and technical solutions in warfare, and as a result Whittle was provided with a factory and extensive support from the Royal Aircraft Establishment.

The story that emerges is very different from the conventional one. Research, development and engineering were marshalled in a new and uniquely British disposition, under Winston Churchill, to wring the utmost war potential out of this tiny island.

Introduction

Early in 1940, Arthur Tedder, a Royal Air Force officer, was taken to see the top-secret Whittle jet engine project. Defeat in France was looming, and Tedder's task at that time was to procure new engines, aircraft and weapons for the next phase of the war. The physical surroundings of the early test were completely different from the well-ordered experimental engine shops of Rolls-Royce or the Bristol aero engine company with which he was familiar, but the improvisation he found at Whittle's company, Power Jets, may well have added to the magical power of the occasion. Tedder found there echoes of the 'crazy inventor' aesthetic that had long been satirised affectionately by British cartoonists, for he recorded his impressions of the test rig as 'pure unadulterated Heath Robinson'. In 'what looked like, and I believe was, a derelict motor garage', he saw what he called 'a typical Emett design' – the prototype of the Whittle engine – and he was astonished, and converted, when he saw, for the first time, the glowing combustion chamber and 'the blazing blue jet flame roaring out into the open'.

Another, almost disastrous, visit by Air Marshal Sir Hugh Dowding, Commander in Chief of Fighter

Command, again underlined the drama of the jet for those who had not seen it before. Whittle showed him the engine and took him round to the outside of the test house where the noise of the jet roar made speech impossible. As they stood about three metres from where the jet nozzle protruded through the test-house wall Whittle pointed to the nozzle, to indicate 'the business end of the engine'. He was aghast as Dowding, practically a god to a middle-ranking Wing Commander like Whittle, misunderstood the gesture and set off briskly almost into the exhaust to be struck by the jet and thrown reeling and staggering across the concrete, his coat torn open and hat rolling away. 'I stood petrified with horror, and when Sir Hugh recovered himself, apparently unhurt, I ... certainly could not speak. ... Until that moment I had not realized how deceptively invisible the jet was to a stranger.'

Like Dowding, Arthur Tedder was a hard-headed airman, and from 1940 he successfully ran the combined air force side of the war in North Africa, playing an important part in the defeat of Rommel, and then becoming General Eisenhower's deputy for Operation Overlord – the D-Day landings and the invasion of German-occupied France. But as with other influential visitors to Power Jets, he felt that he was in the presence of 'a real war winner, justifying the manufacture of an initial batch of engines and aircraft to match, straight off the drawing board.' This kind of 'conversion experience' was not unusual among those who went to see the early jet. It suggests that even when making hugely expensive technological choices – events we might expect to be ruled by cold technical constraint and rationality – we cannot rule out elements of theatre and even of romance.

The Whittle story has a powerful human quality. He showed brilliance, charm and charisma which helped him to recruit major support from government and the RAF for his engine. To the banker L.L. Whyte, 'it was like love at first sight, the impression he made was overwhelming.' Ralph Dudley Williams, Whittle's principle partner in the engine venture, said simply that on meeting him, 'I just fell for him.' However, in late 1943, just as the jet achievement was unveiled in the press ('Britain has fighter with no propeller' was the headline in the *Daily Express*), his company, Power Jets, was nationalised. A colleague called it 'Greek tragedy in the modern world; the hero publicly acclaimed at the very moment when his deeper ambition is frustrated.' The reasons for this drama have never been fully analysed.

Whittle's own account, *Jet*, published in 1953, was, understandably, written in a spirit of bitterness and disappointment. He described at length the difficulties he and the Power Jets team experienced in their dealings with government agencies, and his account of the invention certainly has a compelling quality that resonates with our familiar beliefs about the lot of inventors. Subsequently, virtually all other accounts have followed his view.

The popular histories naturally take an emphatic line on the supposed poor treatment meted out to Whittle, frequently wheeling out the jet story as a supporting element for the familiar assertion that 'Britain is good at inventing but bad at developing.' Whittle's death in 1996 produced a virtual orgy of this kind of comment, with even the most serious newspapers peddling accounts which verged on the absurd and revealed a startling ignorance both of the realities of engineering development and of historical analysis. Some suggested that the jet could have been practical reality years earlier, even in time to give a convincing technical superiority to the RAF in the Battle of Britain. Typical of the genre was the obituary in the *Daily Telegraph* which referred to Whittle 'at times using scrap metal' which intentionally conjured the image of the great inventor rootling for rusty scrap in a car breaker's yard. (Whittle's own statement was simply that 'at least half the engine ought to have been scrapped because of general deterioration.') *The Times* asserted that his ideas were 'scoffed at' by the Air Ministry while the *Guardian* claimed, quite misleadingly, that the Air Ministry 'repeatedly declared ... the idea was largely pie in the sky.'

All these accounts show, or affect, surprising ignorance of the pre-existing background to gas turbine work in several countries. Furthermore, the 'technological determinism' implied by the insistence that a major technological development flows uniquely from a single person's individual genius and persistence (as famously emphasised by Samuel Smiles for James Watt and the other 'heroes' of the Industrial Revolution) is now seen as a less useful way of analysing complex technological changes. Rather few historians, or indeed engineers, given a moment to reflect, would now assert that there would have been no jet engine without Whittle but the obituarist in the *Independent* contended, like those in most of the other newspapers, that 'Whittle changed the lives of countless millions of people throughout the world.' Ascribing a marvellous immutability to the historical account (actually the obituarist's own book) it asserted as fact that 'the Ministry of Aircraft Production did not take the pressure off him and allow him to get on with the job' and that this was 'well documented and part of history'.

It is clear that the notion that Whittle accomplished his engine work against a background of official indifference, or in the teeth of Air Ministry opposition, is so prevalent that it has entered the folklore of the subject both at a scholarly and at a popular level. This view, and the reasons for it, needs to be examined critically.

A study of the jet is also intriguing because of its astonishing effect on world travel and world consciousness. F.T. Marinetti, in a striking anticipation of Marshall McLuhan's notion of the 'global village', wrote, in a typically clamorous 1913 Futurist manifesto, of the dawning transport possibilities. He foresaw 'The earth shrunk by speed. New sense of the world. ... Vast increase of a sense of humanity.' Weighing that notional 'sense of humanity' against the environmental impact and cultural erosion caused by mass travel, or against the global reach of both terrorism and state military action (both equally reliant on the jet), is not the task of this book. But the 'earth shrunk by speed' has now come about, indisputably through the gas turbine engine.

For Britain, particularly, the jet had a special role for it came to form part of the essential underpinnings of a kind of 'Defiant Modernism' – a renewed post-war belief in the role of technology in national life. A replacement 'empire' of high technology with a regenerated industry would sustain the nation and, in this, the story of the invention of the jet, along with the new engines and aircraft themselves, became important symbols of new technique. Furthermore, the jet project became closely bound up with Stafford Cripps' aim, as Minister of Aircraft Production, to modernise the aircraft industry. From 1943, as reconstruction plans got under way, Stafford Cripps argued presciently, and long before terms like 'rust belt' entered the vocabulary, that it would be advanced industries like aviation 'and no longer ... coal and cotton' on which we would now depend. These ideas translated into centralised government direction and generous support for the aircraft and aero engine industries, driven in part by an ambitious policy to rival the USA in civil aviation. This 'Crippsian spirit' infused the post-war Ministry of Supply (MoS) and was carried through by both Labour and Conservative governments to the 1960s when a growing scepticism about the benefits and alarm at the costs of aircraft programmes provoked a reappraisal.

The linkage between jet work and the government began in the 1930s, so it will be necessary to trace the progress of the Whittle jet from the inventor's initial ideas, through pre-war developments, the launch of the Power Jets company and the troubled wartime attempts to build and productionise the engine, to the development of a structure for a post-war gas turbine industry, in which several firms were making and researching gas turbines, supported by the Ministry of Supply.

In all this, the debate is still alive about the way Whittle was treated and the episode raises wider questions about the British management of R&D and its translation into successful business. We will see that the jet programme was both more influential, in terms of providing a new orientation for the UK industry, but also less successful, with respect to the particular progress of the Whittle team, than has generally been appreciated. We will also examine the question that has often been raised as to whether the Air Ministry did too little to assist Whittle. Events in the Whittle saga are still debated and are more than a little mysterious.

The jet engine programme had a complex history, in terms of engineering development, the official administration of the project, the structure of Frank Whittle's firm, Power Jets, and the collaboration between the various firms brought in to manufacture the engine. All these strands suggest, contrary to the accepted view, a considerable degree of flexibility, and indeed originality, on the official side in fostering the new engine and the new industry.

There was the establishment and funding of Power Jets, the company set up in 1936 to develop the Whittle engine, which, it will become clear, was from the outset almost a surrogate official project, and not simply the independent entrepreneurial venture, battling against adversity, that is usually depicted. Then there was the formation of the Gas Turbine Collaboration Committee which became an unusual mechanism for pooling experience across all the aero engine companies and government agencies engaged in the wartime jet engine programme, almost irrespective of commercial rights. The nationalisation of Power Jets in 1944 and its merger with the turbine engine department of the Royal Aircraft Establishment (RAE) to form a government-owned company also suggests a contemporary open-mindedness about exploring new kinds of organisation, and there is finally the conversion of this government firm into a more conventional state research establishment – the National Gas Turbine Establishment (NGTE).

Much of this account is new and draws on a range of interviews over several years, some with key participants, as well as on new archival resources. We will discover that Whittle was treated with considerable indulgence by the Air Ministry and by the Ministry of Aircraft Production (MAP). Fresh sources from the Power Jets side, and official papers, have made it possible to uncover the nature of the wartime relationship between Power Jets and the MAP and the sources of friction. It has also been possible to throw new light on the wartime nationalisation of Power Jets (an event that was particularly resented by Whittle's adherents) and to show that the impetus to take this politically tricky action gained sufficient force because it appeared to satisfy two quite different policy aims. There was, on the one hand, the desire of MAP officials to deal with the stalled jet engine programme and the acrimonious relationship with Power Jets. On the other hand there was the visionary and strategic intention of Stafford Cripps, as Minister for Aircraft Production, to modernise Britain's industrial base and his intention to establish a vibrant aircraft sector sustained by vigorous government R&D establishments. In his scheme Power Jets, with its undoubted talents and brainpower, would become one of these centres.

This book follows the jet from Whittle's first ideas on through the war years, looking at the progress of both the jet team and the engine, and then looks at the de Havilland Comet. The programme to build this jet airliner links closely to the jet 'story' for it was born in the closing stages of the war out of a desire to announce to the world 'the pioneering work on jet propulsion in this country'.

This sense of 'jet modernity' even survived the Comet crashes to sustain the will to build Concorde and to recover 'the leadership we so narrowly missed'. Concorde, therefore, is considered here as a project in intimate connection with the pioneering wartime engine work.

The narrative finally considers the Rolls-Royce RB 211, which emerges as the watershed in the post-war development of the jet, for this engine bankrupted the great national champion during development, but also enabled Rolls-Royce to grow subsequently to become one of the major engine builders in the world, equalling its US rivals in size and market share. Even though it was conceived almost 40 years after Whittle began to promote his engine ideas, the RB 211 still relates to the Whittle story, not least because Whittle himself conceived all the 'mutations' of the jet engine, including this type of bypass or 'fan-jet', but also because the huge initial public commitment to fund this project, through 'launch aid', and to finance completion during the period of receivership, relied on a national belief in the value of the jet – a belief that had been instilled into national consciousness by the epic nature of the story of Whittle and the jet.

In the end pages of the book, some final notes attempt to tease out further some intriguing points: Without the war to spur development, would jet aircraft have arrived in any event? Would they have come later? There is also the wartime development of the jet in Germany to consider, which followed a quite different pattern of control and direction and makes a useful comparison with the development in Britain. And how 'revolutionary' was the jet; is it helpful to analyse it, or to describe it, in this way?

• Chapter 1 •

Whittle's Early Jet Ideas

Frank Whittle entered the Royal Air Force as an aircraft apprentice in 1923. The three-year course on which he enrolled was designed to produce the aircraft mechanics and service personnel required to repair and maintain RAF aircraft. However, by exceptional ability and effort, he was one of only five apprentices (out of 600 in the initial intake) selected to go on to train as an officer cadet and pilot at Cranwell, the RAF training college. Whittle took a keen interest in aeronautical developments and in 1928 his contribution for the cadets' termly thesis was entitled *Future Developments in Aircraft Design*. He anticipated a large improvement in aircraft speed, coupled with an increase in the heights at which aircraft flew, in order to take advantage of reduced air resistance at high altitude. He recognised that in a conventional piston engine power falls off with altitude and considered in some detail, as part of this overall view of aircraft evolution, the efficiency and thermodynamic design of a gas turbine. He observed that although a steam turbine would be impractical for aircraft owing to the weight of boiler and condenser, nevertheless 'the turbine is the most efficient prime mover known [so] it is possible that it will be developed for aircraft, especially if some means of driving [it] by petrol could be devised.'

At this time Whittle considered using the internal combustion gas turbine to drive a propeller. In the following year he realised that a gas turbine could be constructed to produce a propulsive jet. This was an independent insight of his, which transformed the gas turbine problem. This jet engine idea had, however, been anticipated in remarkably modern form by Charles Guillaume, who had taken out a French patent in 1921 – something of which Whittle, as well as other jet pioneers, was not aware. He was also, naturally, unaware of (secret) work by the British research scientist A.A. Griffith at RAE Farnborough, on the idea of a gas turbine to drive a propeller – work which went back to 1926. Moreover, the gas turbine, in a heavy, industrial and relatively inefficient form, already existed and was employed in industrial plants where a cheap supply of combustible gas was available. For example, from 1914 the Thyssen company installed them at several steelworks in Germany where they ran on waste blast furnace gas.

However, the jet propulsion idea made the Whittle gas turbine/jet engine conceptually different from the propeller turbine, in which as much energy as possible is extracted as rotary shaft horsepower from the exhaust by the turbine stages. Whittle's idea instead left as much energy as possible in the gas stream to form a high velocity exhaust jet. This simplification of the gas turbine made Whittle's jet proposal attractive for development at a time when the combined inefficiencies of the compressor, the turbine and the required reduction gearing and propeller drive seemed, in aggregate, too great to make a propeller turbine unit viable.

The basic turbojet consists of three main elements – the compressor, combustion chamber and turbine (see Figure 1 overleaf). Unlike the familiar piston engine, where compression, combustion and expansion (the working stroke) happen repeatedly and in succession, the gas turbine performs these same functions but under conditions of continuous flow.

The problems for such an engine, in the inter-war period, were that all the constituent parts seemed inadequate. First, although high-power compressors were increasingly well understood from supercharging high-power piston aero engines, like the Rolls-Royce 'R' and, later, the Merlin engines, they were not efficient enough and would rob too much power from a pure gas turbine engine. Second, continuous combustion at a high enough rate and in the high airflow required had not been attempted anywhere. Finally, devising a material for the turbine blades, which must stay intact while literally red hot, under huge centrifugal loads and at high rotational

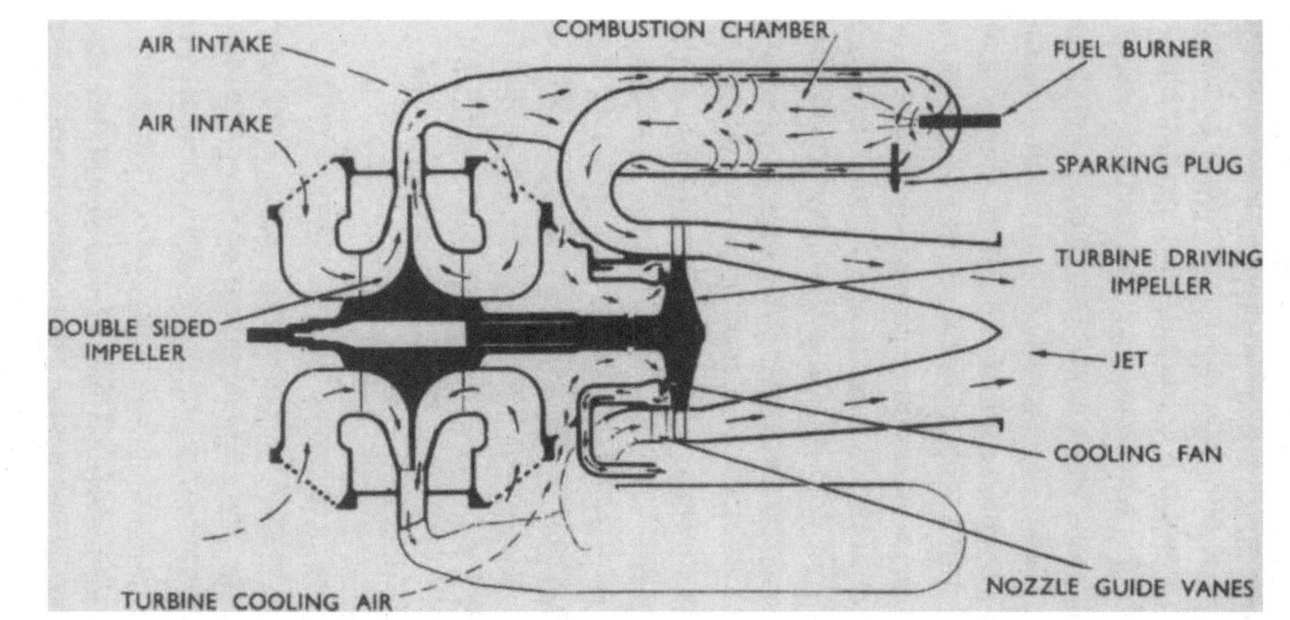

Figure 1: The essential Whittle design. The double-sided compressor (to left) sends air to the double-walled combustion chamber. It reaches the turbine wheel after two 180-degree turns.

speed, was an unsolved conundrum. The exhaust valves of conventional piston engines, it is true, had to 'live' in a flow of combustion gas, but they experienced the worst conditions only intermittently, and for most of the engine cycle were held in the closed position, passing their heat into the cylinder head and engine coolant. In any event, exhaust valve performance was proving at that time to be a limiting factor in piston engine development.

In 1929 Whittle's commanding officer arranged for him to discuss his jet engine ideas at the Air Ministry, where he met W.L. Tweedie, a technical officer in the Department of Scientific and Industrial Research (DSIR), and A.A. Griffith of the RAE. The result, Whittle recorded, was 'depressing' and he subsequently received a letter stating that the engine was impracticable (for the time being) because materials did not then exist capable of withstanding the high temperatures and stresses that would occur in the turbine stage of the engine. The widespread impression of Air Ministry prejudice against the gas turbine at this time derives from Whittle's account of this episode, in which he blamed his reception on 'a very unfavourable report ... that had been written some years before'.

Almost all subsequent authors have repeated Whittle's impression of this report, by W.J. Stern, of the Air Ministry Laboratory, South Kensington, blaming poor Stern for holding back turbojet development by several years. In fact, Stern's report was a professional piece of work which actually anticipated a successful aircraft gas turbine (rather than rejecting it out of hand) but noted that contemporary compressor efficiency was too poor to support a gas turbine cycle while heat-resisting materials for the turbine stage were not yet available. He suggested that 'the internal combustion turbine will not be rendered practical by the revolutionary design of some lucky inventor. The steam turbine engineer and the metallurgist ... are the people with whom the future development of the turbine rests.' Incidentally, Stern, as a member of a special panel of the Aeronautical Research Committee (ARC) in 1930, did recommend construction of a turbine to the design of A.A. Griffith 'if it would provide an unequivocal check on the theory.'

The important point about this episode is that Whittle had been taken seriously enough to be invited to discuss his proposals at high level. Although the letter (probably written by A.A. Griffith) intimated that the Air Ministry did not wish to pursue Whittle's scheme at that time, it observed that 'the internal combustion turbine will almost certainly be developed into a successful engine, but before this can be done the performance of both compressors and turbines will have to be greatly improved. However it has been of real interest to investigate your scheme and I can assure you that any suggestion submitted by people in the

Service is always welcome.' It was, an RAE turbine worker later observed, 'a kind letter'.

Whittle, in his memoir, appears not to realise how exceptional such access must have been for a newly commissioned Pilot Officer: Griffith was then one of the most eminent Air Ministry scientists, a member of the Aeronautical Research Committee and had an important voice, through the ARC's Engine Sub-Committee, in the national direction of aero engine policy. Griffith had also proposed his own gas turbine project (to drive a propeller) as early as 1926, which derived from his new, and highly influential, aerodynamic theory for axial flow turbine and axial flow compressor design. Griffith's paper indicated that the then current axial flow compressors were inefficient because they operated with the blades in a stalled condition. Designing them in the light of aerodynamic theory (treating them, in effect, as rotating wings) would, he argued, allow a large increase in efficiency and make possible a practical gas turbine.

It is an important historical point, usually ignored in conventional jet histories, that from 1937 workers at the RAE (in particular Hayne Constant) began a parallel 'government' gas turbine programme. For speed of results, however, Whittle chose the single-stage centrifugal compressor, since it was relatively well understood from piston aero engine supercharging, while the RAE team concentrated on the theoretically more demanding, and then imperfectly understood, axial flow compressor (see Figure 2, and Plate 3 for a comparison of the two types of compressor). But this project was given less priority than Whittle was to receive, and indeed, some of the most able RAE personnel were seconded to the Whittle programme.

However, the RAE work proved highly influential, and ultimately the main axial flow development of post-war British aero engines was to flow from this work. Griffith's ideas led to a line of transmission through the wartime RAE turbine work, which was the basis of an engine built by Metropolitan-Vickers, the 'F2' or Beryl, to its successor, the post-war Armstrong Siddeley Sapphire, and thence to Avon and the main post-war Rolls-Royce engines. Griffith, in fact, joined Rolls-Royce as Chief Scientist in May 1939. From this perspective, the Whittle engine, with its use of a centrifugal compressor, could be regarded merely as a temporary expedient. Intriguingly, in Germany, Pabst von Ohain, like Whittle an inventor from outside the aero engine mainstream industry, also used a centrifugal compressor for rapid results in association with Heinkel, but the German air ministry directed Junkers and BMW to use axial flow compressors developed in consultation with aerodynamicists at the Aerodynamische Versuchsanstalt (AVA) – in some respects the German equivalent of Farnborough.

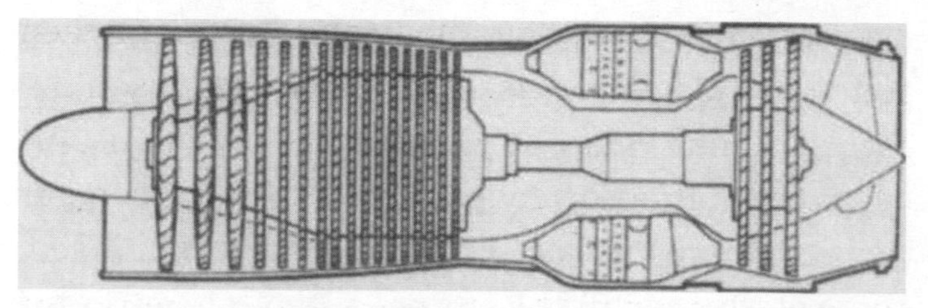

Figure 2: An axial flow engine. This is unlike the basic Whittle centrifugal engine (p. 15), in which a vaned rotor 'slings' the air outwards to accelerate and compress it. Axial flow engines (most aircraft units today) compress the incoming air in stages as it passes through a series of rotors with wing-like rotating blades. (See Plate 3 for further comparison.)

Thus the attitude of the Air Ministry to Whittle can hardly be seen as negative or discouraging. In fact, its actions showed recognition and appreciation of Whittle's aptitude and potential. This can also be seen in the Ministry's decision to allow him to attend Cambridge University. Whittle completed the Officer's Engineering Course in 1933 with distinction. The Air Ministry had, in the past, sent one or two outstanding officers from this course on to Cambridge to take the Mechanical Sciences Tripos but this scheme had been officially terminated in the preceding year. However, Whittle wrote formally asking for special consideration and, in view of his excellent results in the RAF engineering course, the Air Ministry, exceptionally, revived the scheme for him. Additional evidence of sympathetic and favourable treatment can be seen following his First Class Honours in the examinations in June 1936 at Cambridge, when the Air Ministry approved an application from his tutor for him to spend an additional postgraduate year there working with the eminent aerodynamicist Sir Bennett Melvill Jones.

• Chapter 2 •

The Formation of Power Jets Limited

While at Cambridge Whittle received, in May 1935, a letter from a former RAF colleague, Ralph Dudley Williams,[1] c/o General Enterprises Ltd, Callard House, Regent Street (Manufacturers of Genterprise All British Products). Williams wrote:

> *Just a hurried note to tell you that I have just met a man who is a bit of a big noise in an engineering concern and to whom I mentioned your invention of an aeroplane, sans propeller as it were, and who is very interested ... Do give this your earnest consideration and even if you can't you might have something else that is good.*

General Enterprises, the unlikely bridgehead for the 'turbojet revolution', in fact marketed an unglamorous and technically undemanding product – a coin-operated cigarette vending machine – and had been formed by Williams and his partner, J.C.B. Tinling, also a retired RAF officer, with a loan of £1,500 from Williams' sister. Williams recalled that the spur to his contacting Whittle again was a chance meeting at lunch with Tinling's father, a consulting engineer ('very able') who observed, 'there's a war coming – why don't you chaps get into the aircraft business.' (After the war Tinling's father, J.A. Tinling, attempted to claim a share of credit for the development of the jet, writing in 1944 that 'an entirely false impression had been given to the world at large', while his second wife, Daisy Tinling, claimed that 'it should be known and publicly acknowledged that it was primarily my husband's vision and foresight in 1934 which led to the discovery of Whittle ... my step-son was merely an interloper who evidently set out from the start to crib his father's idea and only made a success through his father's financial connections. That may be very clever but it's not cricket ... But for my fancying a lobster at Verrey's [a well-known Regent Street restaurant], on that particular day, ... the Whittle plans would still be in the bottom drawer.'

R.D. Williams had been a fellow cadet with Whittle at the RAF College, Cranwell, in the September 1926 intake. Their batman introduced them, as they were to share 'digs'. Williams, in a striking phrase that echoed the impact Whittle had on many of his associates, recalled later, 'I just fell for him', and that 'I was the person who got on with Frank best.' They remained lifelong friends and, after the war, Whittle even gave an 'eve of poll' address for Williams who was standing as

Conservative parliamentary candidate, although Whittle then was a socialist.

Whittle struck a deal with Williams and Tinling, whereby they would seek commercial backing for the engine and would finance further patents. In return, they were to have each a quarter share of the commercial rights in the engine. They also agreed, at Whittle's insistence, that they would not approach any firms connected with the aircraft industry, because he feared that a big firm might take over the idea, and a patent battle would be too costly to fight. Various approaches to financiers and industrialists failed, until Tinling's father put them in touch with an able consulting engineer and patent agent, Mogens Bramson, who took the engine proposal to the City investment bank O.T. Falk.

Little has been written about the firm of O.T. Falk, but its particular quality of unconventionality, compared with other merchant banks, and the personalities of its members, which included Lancelot Law Whyte and Sir Maurice Bonham Carter, forms a crucial part of the British jet engine story. The founder, Oswald Falk, had been Treasury Delegate to the 1919 Paris Peace Conference and had been described as 'the only high-brow in the city'. Whyte considered Falk personally to be 'one of the Englishmen best informed on the political and military developments in Germany' and the partners and senior employees as 'all exceptionally intelligent men, ethically liberal, and intellectually radical'. The bank was, he believed, 'one of the important nuclei of anti-Hitler and pro-Churchill opinion in London at that time', and it derived additional impetus for this stance since, it was said, Violet Bonham Carter, one of the most influential anti-appeasement campaigners (and wife of Maurice), was conducting a *menage à trois* with her husband and 'Foxy' Falk.

Lancelot Law Whyte was himself an unusual figure – a philosopher and an intellectual who had worked in, and kept up with, theoretical physics and who had a powerful interest in and sense of historical process. After Cambridge, where he worked for a time in the emerging field of atomic physics in Rutherford's laboratory, he travelled to Göttingen in 1924, becoming friendly with Max Born and hearing Neils Bohr lecture on the new theory of the atom. Subsequently, in Germany, he met and had discussions with Einstein before deciding to leave academic life. He entered merchant banking through the mediation of Montagu Norman, Governor of the Bank of England. Norman sent him to see Sir Maurice Bonham Carter, a partner at O.T. Falk. According to Whyte, Norman broke his rule not to use his influence in City appointments because Whyte's sister and Norman's mother were both Christian Scientists. The City appealed to Whyte intellectually because 'in the City one saw human desires being expressed in quantitative form. ... Did stock market prices quantify human lusts in the same way as the clinical thermometer converted human pathology into a numerical temperature?'

Through his scientific training Whyte became the member of the bank most concerned with venture capital for proposals with a scientific or technical element, 'raising money for the rare deserving cases'. Thus he became a director in the early 1930s of Scophony, the company established to exploit John Logie Baird's television system, but nursed 'a secret hope ... that something wonderful would turn up for which I would throw over everything else.' And although at first reluctant to meet, when told by Bonham Carter that 'a young flight lieutenant had invented a new aero-engine', he found this in Whittle. He wrote:

> *It was like love at first sight, the impression he made was overwhelming. I have never been so quickly convinced, nor so happy to find one's highest standards met. Whittle held all the winning cards: imagination, ability, enthusiasm, determination, respect for science, and practical experience – all at the service of a stunningly simple idea: 2,000 h.p. with one moving part. This was genius, not talent. ... That night I told my wife that I had met one of the great inventive*

> *engineers of our time. ... it was like what I imagined was the experience of meeting a saint in a much earlier religious epoch: one surrendered to the enchantment of a single-minded personality born to a great task.*

Whyte's background in physics, his semi-mystical personality and his historical overview of the subject were crucial in conditioning his response to the Whittle jet. His reaction was more that of a historian of science and a physicist than that of a banker for, in evaluating it, he transferred directly to aero engineering the concept of 'elegance' which so often is advanced as a touchstone for the evaluation of mathematical proofs and scientific theories. 'The elegance of the idea won me. Every great advance replaces traditional complexities by a new simplicity. Here it was in the iron world of engineering.'

Armed with a favourable report from Bramson, whom the bank had now formally asked to analyse the Whittle scheme, Whyte set out to raise capital and contacted Sir Henry Tizard for a supporting opinion. Tizard was, at that time, an important figure in the Aeronautical Research Committee, the body responsible for the overall direction and evaluation of research work done in the government research establishments, such as RAE Farnborough, and within the firms on Air Ministry contracts. He also chaired the Engine Sub-Committee of the ARC. Tizard was a scientist who was uniquely trusted by the Air Ministry and the RAF. This reliance on him resulted, in part, from his work in the early development of the technical and scientific testing of military aircraft in the First World War, when he came from Oxford as a research chemist, learned to fly and began systematic testing of aircraft performance and behaviour at the Martlesham experimental station in Suffolk.

However, the respect in which he was held also derived from his wide scientific understanding and his highly incisive and pragmatic judgement. In the period immediately after the First World War he did important pioneering work with the most notable independent British piston engine researcher and consultant (and influential member of the ARC Engine Sub-Committee), Harry Ricardo, and with David Pye (subsequently Director of Scientific Research at the Air Ministry) on the 'detonation' properties of fuels, leading to an understanding of how fuel quality limited piston engine output, and anticipating the standard 'octane rating' for petrol/gasoline that was established a few years later in the USA.

At the time of Whyte's approach Tizard was deeply immersed in the development of the revived air defence system for the UK and in the debate about the possibility of German bombers delivering 'a knock-out blow' to Britain. The Committee for the Scientific Survey of Air Defence (more usually known as the Tizard Committee) had first met in January 1935 and, from the outset, became the nursemaid for the emerging technique of radar. Tizard personally was central to the promotion of radar research and its implementation to create the radar 'chain' around the eastern and southern coasts of Britain. However, he showed a striking strategic vision which went far beyond the actual technique of radar, for he realised that in order for it to become a powerful addition to air defence it had to be integrated into the operational control of aircraft. Thus even before radar was available, he asked for a series of experiments to be performed by the RAF (the Biggin Hill Interception Trials) in which controllers directed defending fighters by radio onto 'dummy' intruders flying a known track, in order to explore the technique of interception and the 'vectoring' of the defenders.

These trials enabled the crucial new techniques for communication and ground control of fighters to be developed, together with procedures for reporting, plotting and directing them to the intruders. Most importantly, by starting from this perception of Tizard's that 'the interception problem was different to the detection problem' the Biggin Hill Interception Trials assessed and developed the accuracy with which fighters could be set onto intruding aircraft and were absolutely crucial to the development of fighter control and the RAF war-fighting system.

The trials also indicated an alteration of emphasis for engine development. Before radar, defending aircraft (in which little reliance was actually placed) were expected to spend considerable time ready at height in 'standing patrols' and reasonable fuel consumption was needed. The Biggin Hill trials showed that the emphasis should change towards very high-power engines, giving the maximum speed and rate of climb from take-off on receipt of the radar warning, even at the expense of heavy fuel consumption, and Tizard argued explicitly from 1935 for 'the maximum excess speed over the bomber'. Privately, he wrote to his friend Harry Ricardo that 'what I want is an engine which gives a terrific power for its size and weight ... through a high consumption so that long distance bombing machines could not compete.'

The Whittle engine clearly fell into this category, and this tactical defence thinking lay behind the encouraging terms in which Tizard wrote to Whyte: 'I am particularly interested in this work because I think that, if we are to provide the high powers which will be necessary for the aircraft of the future, we must develop some kind of turbine.' It seems clear that it was only because the jet (which had not yet run) appeared to fit closely into this emerging tactical requirement, revealed by the interception trials that Tizard had asked for and studied so carefully, that it was considered at all, given the high level of demand on both the R&D and the productive capacity of the industry imposed by the RAF expansion programme. Tizard added as postscript to his letter: 'Of course I do not mean to imply that success is certain. All new schemes of this kind must be regarded as "gambles" in the initial stages. I do think however that this is a better gamble than many I know of, on which much money has been spent!'

Tizard also noted that he had a very high opinion of Whittle, who had 'the ability, energy and enthusiasm for work of this nature' and an intimate knowledge of practical conditions. 'This combination of qualities is rare', he wrote, 'and deserves the utmost encouragement.' Tizard rated it highly probable, given adequate finance, that Whittle would succeed in producing the new powerplant.

O.T. Falk put great store on this letter, quoting it to possible investors. Another point that Tizard made also confirms the linkage with rearmament: 'My general opinion of the importance of this work leads me to express the hope that the money will be raised privately so that the knowledge that it is going on will not be widespread.' This and other observations suggest that the capitalisation of Power Jets could be regarded almost as a surrogate official venture, and certainly as an officially sanctioned one. A letter drafted by Maurice Bonham Carter to Lord Wakefield, head of the Castrol petroleum company and a well-known promoter of aviation, summarised the thinking behind this position, which appears to have been tacitly accepted both within the Air Ministry and in Power Jets:

> *It is generally agreed ... that at this stage, both to avoid delay and the restrictions necessarily imposed by finance from the Treasury, the initial expenditure should come from private sources ... I am approaching a very few private individuals only who recognise the nature of the business and its public interest.*

In this context, and within this circle, the phrase 'public interest', I suggest, signified air rearmament. To the banker Peter Samuel, Bonham Carter made more explicit the quasi-official support for the project and the understanding, among those responsible for finding finance for Power Jets, that public money would make an essential contribution. He noted that:

> *It is of course accepted by all concerned that the development of such an engine to the stage of commercial production requires a large expenditure far beyond any sum which we have in mind to*

raise now. We anticipate that material contributions towards this expenditure will be provided in due course from official sources by way of development grants. As you are aware, we are already in negotiation for an initial research grant. But it is agreed by all concerned (including, as you will see from his letter, Sir Henry Tizard) that in order to avoid the delay and the restrictions imposed by finance from the Treasury, the initial expenditure should be raised from private sources.

Bonham Carter also stressed that the company's expenditure was being conducted 'on a very conservative basis', that the Board was receiving no fees and that 'Flt. Lt. Whittle, who is seconded for service for this work, is remunerated by his normal pay as an officer in the Air Force.' The new company was registered on 19 March 1936 and the basis of its constitution was the so-called Four-Party Agreement between O.T. Falk, the Air Ministry, Whittle and, together, Williams and Tinling. O.T. Falk was represented on the board of the new company by L.L. Whyte (as chairman) and Sir Maurice Bonham Carter as a director. The other directors were R.D. Williams and J.C.B. Tinling, while the Air Ministry contributed Whittle as Chief Engineer at no cost to the company for a period of five years.

Whittle, in *Jet*, and almost all subsequent authors have made much play with the arrangement set out in the Four-Party Agreement whereby Whittle could act as honorary Chief Engineer to the new company 'provided that the work ... shall not conflict with his official duties and ... shall not ... in any one week exceed a total of six hours.' This is disingenuous. The Air Ministry may not have wished to openly admit that Whittle was assigned full-time to a privately financed company for the purposes of developing an engine that many regarded as fanciful, but that clearly was what occurred. During his extra postgraduate year at Cambridge up to June 1937, which the Air Ministry had approved and financed, Whittle worked largely on his engine design. After this he was placed on the 'Special Duty' list and was not assigned to a squadron or to an RAF station. These measures make it clear that there was tacit acceptance of his central role at Power Jets and this, in itself, was a substantial contribution – a pilot who was acknowledged to be of the highest quality, trained at substantial public expense, was, at a time of deep foreboding about national defence, contributed to a rather risky engine development programme. As further evidence of exceptionally favourable treatment, in this period Whittle was promoted to the rank of Squadron Leader, for which he was excused the usual examination.

We can see that from the outset the structure put in place to finance the Whittle jet was odd. It is true that it initially used private capital, although the single most substantial investment was from Lord Weir, who was prepared to risk funds in the national interest. Weir had been Controller of Aeronautical Supplies at the Ministry of Munitions in the First World War, where he had been assisted by Sir Maurice Bonham Carter, and a firm friendship between them dated from that time. By May 1935 Weir had returned to munitions production, joining the Air Ministry as a special adviser to the Secretary of State and refusing to accept payment for this public service. However, he took up shares in Power Jets privately through his Glasgow engineering company G. & J. Weir following a direct approach from Bonham Carter.

Power Jets and the Air Ministry now began negotiations to establish a form of development contract whereby the Ministry would pay progressively for research and running experience with the engine. David Pye, as Deputy Director of Scientific Research (DDSR) at the Air Ministry with a special responsibility for engine development, proposed a series of staged payments to Whyte in July 1937 amounting to £5,000. After twenty hours running and when the engine had reached a designated speed, the Air Ministry was to purchase the unit for a further £5,000 but would put it at the disposal of Power Jets for further running. This schedule was arrived at by Pye in discussion with his assistant, William Farren.

Both these men have been depicted by Whittle as being sceptical about jet propulsion, and as obstacles to him, but we should perhaps take Pye's letter to Whyte, as Chairman at Power Jets, at face value. He wrote: 'My only concern is to devise some kind of co-operation which would be financially acceptable to the Air Ministry, and would ensure that research proceeds actively with the present unit.'

Pye also added that the series of contracts he proposed was 'quite outside the normal run of such things [but] the whole project is exceptional and calls for exceptional treatment and if an arrangement on these lines would be acceptable to you I will do my best to see it through. It leaves no room for doubt that we at the Air Ministry regard the scheme ... as theoretically sound.' The schedule proposed was:

1. £1,000 for a full report on all work to end of July, running unit up to 12,000 rpm.
2. £2,000 for 10 hours further observed running up to 14,000 rpm.
3. £2,000 for 10 hours at speeds up to 18,000 rpm.
4. On completion of this research running, £5,000 to be paid for the unit, with the intention to put it at your [Power Jets'] disposal for further running.
5. Separate research contracts for associated work such as experimental combustion chamber tests.

In fact the sum offered was subsequently reduced to £5,000, possibly on the advice of the Ministry's Contracts Directorate, which appeared to be uneasy about the firm's standing and its close relationship to a merchant bank. At this time the share capital raised by O.T. Falk and Partners was only of the order of £5,000 and, furthermore, Falks did not take up its option to extend its own shareholding to £20,000. Even Whittle's sympathetic biographer, John Golley, has had to admit the quandary in which this placed officials. 'The Air Ministry ... wished to make an offer to the Company which would be sufficient to encourage further finance to back the project, but they mistrusted a group of financiers who began to get cold feet, after raising only £5,000.' The official history also makes the point that 'in the past, as a matter of principle and convenience the Air Ministry had given financial assistance only to the well established firms in the aircraft and engine industry and did not give financial backing to bankers, investment houses or promoters, no matter how close their connection with the aircraft industry. And it so happened that ... Whittle was being sponsored by a City firm, and financial assistance to them would have been a new departure and a precedent.'

Although these sums may now seem small, the proposed initial government contributions amounting to £10,000 equate to almost £500,000 at 2003 values and should be related to the total authorised share capital for Power Jets of £25,000. In relation to the scale of the company's operations these were substantial payments. However, they were not adequate to sustain the experimental programme and it is fair to note that Air Ministry officials did have some misgivings about Power Jets. Certainly the capabilities of the new firm were meagre, compared with the companies they usually dealt with like Rolls-Royce and Bristol which had huge numbers of operatives, draughtsmen, design engineers and so on. In 1938 Whyte approached the Air Ministry for further funding. The official history notes that:

> *The Air Ministry did not rate the judgement or the resources of the firm very highly. The Director of Scientific Research had early expressed the fear that the Directors of Power Jets were over-optimistic about the speed with which results would be obtained. When Power Jets began to ask for help at the first hint of development difficulties, which were no greater than those which experienced engineering firms would have considered inevitable and taken in their stride, the authorities in the Air Ministry felt confirmed in their low opinion of Power Jets.*

We have seen that the constitution of Power Jets was unusual. There was the low-key Air Ministry contribution of Whittle, with partial official funding and encouragement. These were elements that were, with hindsight, not ideal, and which may have helped to sow the seeds of trouble later, but it is hard to see how a fully funded experimental jet programme could have been created, unless the Air Ministry had insisted on the development programme being placed with one of the Ministry's trusted contractors, such as Rolls-Royce. It is to the credit of officials that they took the risk of proceeding with Power Jets at all and it seems clear that the basis on which the company was established was tacitly understood by both sides. Robert Schlaifer, who wrote a masterly history of aero engine development in the immediate post-war period, and who had the advantage of communication with Sir Maurice Bonham Carter, noted that 'it was realised by Falk and Partners from the beginning that the entire undertaking would ultimately have to be a partnership with the state, and the extensive rights granted to Whittle were granted to him in large part as a representative of the state.' The closeness of the enterprise to British rearmament efforts was also underlined by the understanding that Power Jets would abstain from approaching foreign sources for funds.

By 1938 a variety of high-power engine work was being conducted in Britain in response to the bomber threat. The Whittle jet was at one, more speculative, end of the 'portfolio' of R&D investments sponsored by the Air Ministry, and there was also an axial flow jet engine emerging at the RAE under the auspices of Griffith's 'disciple' Hayne Constant. Tizard had a great sense for energy in people and projects and also sought to promote the rival 'government' gas turbine at the RAE being built in association with Metropolitan-Vickers, but he noted that the company 'was making progress but only slowly. There was no real drive behind it.' By contrast he did sense this drive in Whittle. Unfortunately, his influence in engine developments, and defence science in general, waned during the war as a result of the well-known antipathy between him and Churchill's preferred scientific adviser, Frederick Lindemann. This was a tragic loss in many fields, but the Whittle project in particular might well have benefited from Tizard's tremendous pragmatism and good sense, both for engineering development and for professional arrangements, if he had retained greater influence. Nevertheless, he did have a continuing role as a scientific adviser to the Air Ministry, and certainly made a useful contribution to the project, not least by widening the number of firms involved and bringing the well-established British aero engine designer, Frank Halford, and the de Havilland company into jet work.

Radical and ambitious piston engine projects

were also under way, including the Napier Sabre, a massively complex 24-cylinder H-pattern engine designed by Frank Halford intended ultimately to give 3,000 hp, and the Rolls-Royce Crecy, a novel 12-cylinder two-stroke designed by Harry Ricardo and enthusiastically promoted by Sir Henry Tizard as a 'sprint' engine for interceptors. Finally, there was the 'blue chip' investment, the relatively conventional but highly optimised V-12 cylinder Rolls-Royce Merlin, then approaching 1,000 hp as it was developed to take advantage of the newly available 100 octane fuel which Tizard and others worked hard to source in adequate quantities.

It is important, therefore, to see the Air Ministry's support for the Whittle engine as one element in a strategic initiative in the engine field, which itself was part of the larger programme for the expansion and re-equipment of the Royal Air Force from 1935 onwards. In the context of this rearmament the Whittle engine was allocated a quite reasonable share of resources, given that, by 1935, it was considered by the Air Ministry that production for rearmament had risen to 'the utmost capacity of aircraft firms'. In these few years the RAF moved from biplane fighters with engines of 500 hp, capable of 200 mph, carrying two machine guns and weighing about 3,000 lb (1,360 kg), to a new generation of equipment – the new Spitfire and Hurricane monoplane fighters with 1,000 hp, eight machine guns, a 330 mph top speed and a weight of 6,000 lb (2,724 kg).

In 1929 the total strength of the RAF was under 29,000 people. This force was expanded more than threefold over the next three years, to over 90,000, with progressive increases in every subsequent year. When it is considered that the majority of RAF personnel required a considerable degree of technical training, while at the same time the technical complexity and capability of the equipment was increasing enormously, the expansion must be seen as an extraordinary achievement.

In addition to new designs and re-equipment, as we have seen, entirely new techniques for the operation and control of these aircraft had to be developed. This re-equipment and training programme in itself constituted a revolution in equipment and tactics for the RAF and it must be acknowledged as a brilliant success. The force that was created proved to be just adequate, during the Battle of Britain, to hold the Luftwaffe, which had been established specifically to wage offensive war by one of the most technologically advanced and industrially competent nations on earth. Considering the urgent needs of this radical re-equipment and retraining of the RAF it is easy to understand why more resources were not devoted to the Whittle jet or to any other long-range piece of weapons research.

In spite of the rhetoric of the jet, and what we now know of its capability, the Merlin-engined Hurricane and Spitfire could not remotely be considered as obsolescent weapons by the time war broke out. In fact, the Merlin engine was the very pinnacle of mechanical engineering achievement, pushing contemporary manufacturing and materials science to the utmost, so that as late as 1937 Ernest Hives, the head of Rolls-Royce, was still describing Merlin development and production problems as 'a nightmare'. It was a nightmare that Rolls-Royce overcame by their resolute engineering approach, but the Merlin programme demonstrates, as Air Ministry production planners knew well, that, in all Second World War aircraft developments, the engine was at the core, with the longest lead time and the most challenging technology.

As RAF rearmament (coyly referred to officially as 'expansion schemes' to appease the appeasers) gathered pace, a vastly increased provision for the production of these new and highly complex aero engines had to be set in train. Thus 'Shadow' factories were developed, partnering, for example, the Bristol aero engine concern with car manufacturers including Austin, Rover and the Rootes group (owners of Humber, Hillman and Sunbeam). Each Shadow factory had to be fully equipped with machine tools, gauges, jigs and so on identical to those used in the parent concern, for 'Shadow' engines required the most faithful replication of all the processes which the original manufacturer had developed so painstakingly. Any deviation in machining technique, materials treatment and so on could produce weaknesses or raised stresses, and at first the car manufacturers were sceptical of the elaborate techniques, having so successfully found out how to achieve mass production for cars economically. However, the 'car barons' soon came to understand that aero engines were special, spending much of their life producing their full design power, and with very high heat flows running through critical parts, whereas popular cars (especially then) used full power only occasionally. The Bristol Shadow Group was chaired by Lord Austin, founder of the Austin car company, then 70 years old, but 'fully in command and admirably detached in handling this unique band of otherwise cut-throat rivals, keeping the discussion always to the point of co-ordination in the country's interest.'

Rolls-Royce were chary of the Shadow scheme, preferring to manage their own production, and two new factories, at Crewe in Cheshire, and Hillingdon, outside Glasgow, were built at government expense, although subsequently Merlin production was licensed to Ford in England and to Packard in the USA.

Driving all this along at the most critical time was Wilfrid Freeman, a brilliant RAF officer, 'full of élan, and the spirit of adventure, a visionary, ... a Pucklike character ... sometimes of almost feminine intuition.' Those working with Freeman included Arthur Tedder, then Director General of Research and Development, and George Bulman as Director of Engine Development and Production.

In the same way that Tizard drove the radar programme, through his high connections and great reputation, so too did Freeman establish the production basis on which British survival was to depend. As with radar, where Treasury understanding of the dawning international crisis led to the bypassing of normal rules, the Shadow scheme often ran ahead with verbal agreements. In 1938, when Ford undertook to build Merlins, Freeman sent George Bulman to the Treasury in person to get verbal authority for £5 million.

Bulman, who is important to this story, had become involved in aero engine development in the First World War. He entered the War Office in 1915 as a young engineer in the Aeronautical Inspection Department and was intimately involved in maintaining quality of production. Spending part of the time in France, in close contact with the fighting squadrons, and also liaising at home with the engine firms, he was acutely aware of the importance of engine quality and reliability for the lives and the fighting ability of the aircrew.

In the inter-war period Bulman, determined to

hang on to the maximum engineering capability in the sector, carefully husbanded the drastically reduced Air Ministry funds for engine development to support projects that would keep what he regarded as the really competent firms – Rolls-Royce, Bristols and Napier – commercially interested in military aero engines. He saw this as an almost sacred duty, describing his role as 'midwife' for new engines.

> *Working in closest partnership with the engine firms one had to decide at what point … to bring in a fresh type, … when to discard the established engine of high repute for one of higher performance with a certainty of a period of teething trouble and criticism; how best to build up and maintain output while keeping operational quality as the vital essential; to decide whether a new defect in service was an odd freak, or heralded an epidemic, demanding perhaps drastic action to avoid … a hold-up all down the production line.*

This personal history, and this realism, partly explains Bulman's scepticism of Whittle and the Power Jets firm, particularly as the inventor had explicitly stated that 'in no circumstances would we talk to anyone connected with the aircraft industry' – the highly expert firms which Bulman had been at such pains to nurture through the 'lean years'.

Bulman refused to be overwhelmed by the potential, or the romance, of the jet, commenting later that 'the better is so often the enemy of the good, and must be proved to be, in all respects, not less than the good.'

It is, possibly, a mistake to personalise this movement in rearmament too much. There were, after all, strong formal procedures for the expansion of air defence. The ambitions of the expansion schemes grew from the provision of 3,800 aircraft over two years in 1936 to 12,000 by April 1938. Similarly the total RAF budget (including research, procurement and operations) went from £17.5 million in 1934–5 to £74.5 million in 1938–9. But the records and reminiscences show that there were public servants in crucial positions who were utterly committed to the defence programme and who pushed the policy along through autonomous efforts. For radar, the novelist C.P. Snow, who worked during the war on the allocation of scientific manpower, commented that 'two successive secretaries of the Cabinet, Hankey and Bridges, did much more than their official duty in pushing the project through.' So too with the expansion schemes. One Sunday in September 1938, Freeman returned to his staff from a meeting in Whitehall with the news that Neville Chamberlain had come back from Germany with Hitler's commitment to peace. 'Peace in our time they say. That means we may have a few more months to get ready. How can we increase our programme?'

The battle for France finally ended in June 1940, and the Battle of Britain, the German attempt to destroy British air power as a prelude to invasion, began, according to Air Marshal Dowding at the head of Fighter Command, on 10 July. In the aftermath of the British evacuation at Dunkirk, Fighter Command had left 768 fighters, of which just over 500 were fully serviceable. By early August, just before the full weight of the German attack, heroic production efforts had built the fighter numbers at the squadrons up to just over 1,000 Spitfires and Hurricanes with over 400 in storage units ready for despatch. Against this the Luftwaffe had a similar number of Messerschmitt 109 single-seat fighters, but also over 200 Messerschmitt Me-110 twin-engined fighters and over 1,000 bombers and other aircraft.

Day after day, the German attacks persisted, sometimes in small groups, sometimes in huge waves at various heights, leading one British pilot climbing for attack to describe the sight as like standing at the bottom of the escalator at the Piccadilly Circus underground station, looking up. On 15 September, the turning point, when German sources believed British fighter numbers to have been drastically weakened, the Luftwaffe suffered one of its worst days of attrition. In the worst week for Britain, at the end of August, some 150 Hurricanes and Spitfires were lost but, through the production programme, stimulated by Churchill's old friend the press magnate Lord Beaverbrook, now installed as Minister for Aircraft Production, the RAF managed to hold on to an almost constant level of aircraft and pilots throughout the battle.

The battle proved that, technologically and numerically, the force was just good enough; there were just enough. Survival had relied on all the pre-war investment in development, production and intense programme planning, followed by Beaverbrook's desperate short-termism (no more production planning for the duration of battle – just 'all that can be done'), to squeeze the utmost fighter production out of the system Freeman and others had set up.

Predicting engineering development at the limits of knowledge is an art, informed by experience, but always a gamble. Men like Dowding, Tizard, Freeman and Bulman had made many correct judgements in building the air defence system. The jet was long-range research; how fortunate that they had not been seduced into a greater commitment to it by an imagined 'jet modernity' and by that 'blazing blue jet flame roaring out into the open'.

• Chapter 3 •

Wartime Development

The Air Ministry has often been condemned for not supporting Whittle more. As we have seen, this is unfair: the Air Ministry did find ways to facilitate Whittle's work, using an unconventional route, and came, by 1939, to support the work almost entirely from public funds. But should the Ministry have done what it did in 1939 earlier – perhaps in 1935, or even in 1929? The essential point to note about the jet engine, which events were to prove, was that, even in 1939, its development was premature. By 1939 the metallurgy, the techniques for machining complex shapes and for fabricating the new components needed for the engine, instruments for vibration measurement and even the theoretical tools for analysing airflow through a jet engine, were barely adequate for the task of creating a functioning engine. Had it not been for the urgent expansion of the Royal Air Force and the attention to British fighter defence capability, there would have been little rational reason to force development so far from what was known. The troubled wartime development of the Whittle jet suggests that this was not an invention whose natural time had come. Since the Air Ministry has been subjected to such persistent attacks, it is useful to observe that impeccable engineering authorities were equally sceptical. When the visionary official in the German air ministry, Helmut Schelp, who did most to launch the German turbine programme, attempted to recruit Daimler-Benz for gas turbine work in 1938 Fritz Nallinger, the head of development, argued that though the turbine might one day be of use the time was not ripe and declined to do any work on it.

Since the story of the British jet engine is, in the early- and mid-war years, one of conflict and disappointment, we need to consider whether the causes of this conflict can be teased out beyond the simplistic identification of 'official disinterest and political manipulation' and 'the apathetic malaise of industry and government' – opinions that have already been touched on. For this it is necessary to follow, in some detail, the steps that were taken to convert the prototype into a production engine for service use.

The new Power Jets company had no production facilities of its own and so had entered into an agreement with the steam turbine specialists British Thomson-Houston at Rugby for detailed design drawings and the manufacture of an experimental prototype. Whittle and the small number of Power Jets personnel also took up residence at the

BTH factory and began conducting tests there from October 1936. However, after some frightening incidents during test runs, Power Jets were moved to a former BTH foundry nearby at Lutterworth in Leicestershire. The engine went through two redesigns and, in October 1938, testing began on the third version which defined, in its general architecture, the form the Whittle jet engine was to take during the period that the inventor remained in control.

Successful results began to accumulate, and by June 1939 David Pye (now Director of Scientific Research) witnessed a twenty-minute test run at speeds of up to 16,000 rpm – an experience which, in Whittle's opinion, marked 'a dramatic change in D.S.R.'s attitude'. Meanwhile Sir Henry Tizard, who was present at a trial in January 1940, remarked, with an insouciance that he may have come to regret, that 'a demonstration which does not break down in my presence is a production job.' As we have seen, Air Vice Marshal Tedder, as DGRD (Director General of Research and Development) at the Air Ministry, also saw a test run and felt he was in the presence of 'a real war winner', justifying the risky step of committing to production and short-circuiting the usual process of prototyping and development.

The ground test engine (known as the Whittle Unit or W.U.) then served as a model for a geometrically similar unit, the W.1, which was built from new components, developed and certificated for initial trials of the Gloster-Whittle E.28/39 aircraft. These trials were strikingly successful. In particular, the engine was regarded as remarkably trouble-free considering the entirely new principle of operation. However, the E.28/39, although originally planned as a fighter, was not suited for military development since the thrust of the W.1 engine was inadequate to allow a reasonable load of fuel and armament. It was therefore built (with a second example) as a pure jet engine research and test aircraft.

For military needs the design was put in hand in 1940 with the Gloster company for a twin-engined fighter aircraft, the F.9/40 (subsequently known as the Gloster Meteor), while a design contract was also placed with Power Jets for an enlarged and more powerful (but architecturally similar) version of the engine, known as the W.2, intended for this aircraft – a development suggested by Whittle. Many of the ensuing problems of the Whittle engine programme stemmed from this apparently unproblematic decision to enlarge the engine. The W.2 therefore became a new design, for which no direct test experience existed.

The challenge of this engineering development at the limits of what was possible cannot be overstated. The enlarged impeller for the W.2, being heavier and more highly stressed, became liable to fatigue and burst. In addition, the gas velocity at the compressor outlet became supersonic on occasion, producing a shock wave which could destroy the impeller. Added to this was the newly emerging phenomenon of 'surging', in which the gas flow in the engine becomes unstable, even possibly reversing direction, sounding like a series of explosions, and setting a limit to the flow and thrust. The engine had also acquired a destructive mechanical resonance period at critical speeds. These were all subtle problems which could not have been predicted in 1939. Impeller failures were frequently catastrophic, since the debris would pass through the engine and into the turbine, but so too were many other types of failure, necessitating frequent complete rebuilds of the test engines.

The extraordinary demands, mental and physical, involved in this struggle with the jet are almost impossible to imagine now. Sometimes the test engines ran sweetly, demonstrating the huge step-change into the new world of vibrationless high-speed flight that the inventor foresaw. Sometimes, with a redesigned and supposedly improved component, or at different speeds and throttle settings, they were baffling and frighteningly unruly, showing the new problems for mechanical design, internal aerodynamics, combustion and high temperature metallurgy that remained to be solved. All these problems, way beyond what had been met before, seem in part to have created a special mentality in the Power Jets team. By day and by night, working 80-hour weeks, they were confronting new, almost intractable problems, solving them by immense effort and resolution.

For example, Whittle recorded that one W.2 engine was 'dismantled, modified, reassembled and tested' five times in ten days – an extraordinary load for this compact small team. He and his colleagues believed that they were about to give Britain a weapon of new and transforming power and so the struggle to perfect the engine was a personal and emotional crusade. To Tizard he wrote:

> *The responsibility that rests on my shoulders is very heavy indeed ... either we place a powerful new weapon in the hands of the Royal Air Force or, if we fail to get our results in time, we may have falsely raised hopes and caused action to be taken which may deprive the Royal Air Force of hundreds of [conventional] aircraft that it badly needs .... I have a good crowd round me. They are all working like slaves, so much so, that there is a risk of mistakes through physical and mental fatigue.*

L.L. Whyte recalled the nervous tension of the time at the height of the war. 'From 1939 to 1941 the tension was never relaxed for a moment. We would leave board or committee meetings after exhausting battles with civil servants or with collaborating firms to return at once to engine tests at Lutterworth … and frequently finish with night duty listening to the German bombers, as for example on the night of the first mass raid on Coventry, only twelve miles away … and this against the fall of France, the rally under Churchill, the Battle of Britain, and the risks of an invasion.' The Power Jets engineer G.B.R. Fielden, who joined the project straight from Cambridge, recalled '80 hour weeks, … no girlfriends, … no one knew whether we bombed, over-run …, the black out, sleepless nights, the fires of Coventry burning … everyone at a pitch.'

It is hard not to sense that this shared achievement in a small team, wrestling with truculent natural forces, translated into a kind of intransigence in dealing with outsiders, whether Ministry men or industrialists working on adapting the engine for production. Perhaps this would not have mattered if Power Jets had been a kind of 'Skunk Works' or advanced projects outfit within a larger firm which could have mediated these outside contacts, provided advice and support and fielded alternative advocates to meet government officials. Power Jets, by contrast, was relatively isolated in Leicestershire, battling with natural forces and then too with human ones. The result was what one Power Jets man recalled as 'a general xenophobic feeling at Power Jets'.

• Chapter 4 •

Even With All the Mistakes – Human and Otherwise

Power Jets did not have the production resources to build aero engines in the quantity required for the RAF. Moreover, its relationship with the BTH company was deteriorating and the jet men were coming increasingly to suspect that BTH was attempting to appropriate the jet engine as its own product. Whittle also became involved in a dispute about the correct form of turbine blading for the new jet ('vortex' blading) which was a change from steam turbine practice. Although BTH did, in fact, agree to make blades in accordance with angles specified by Whittle, he felt that 'the affair more or less left a permanent scar on the relationship between Power Jets and the B.T-H'. It is hard to understand why this incident provoked such a 'scar' except perhaps in the light of Lord Kings Norton's assessment of Whittle. Kings Norton (then Harold Roxbee Cox), a Farnborough government scientist, was brought in mid-way through the war as 'Director, Special Projects' (code for the jet) to liaise between Power Jets and the rest of MAP and between Whittle and the other engine companies.

He declared: 'I had the highest opinion of him ... a brilliant leader ... [but] he was a curious mixture .... My sympathies were with him .... Not a tactful man ... [but] had he been more reasonable he wouldn't have got so far ... no sense of humour whatsoever.'[2]

Apart from the question of the relationship, BTH had no experience of aero engine practice and so Whittle and Air Ministry officials looked for a technically competent engineering firm to build the jet and the Rover car company appeared to be an attractive candidate. It had a reputation for building good quality cars and could be regarded as an 'engineering-led' rather than a 'cost-led' company and therefore suitable for a new type of job like this. It was involved in the aero piston engine 'shadow' production scheme and the engineering direction of the company was in the hands of Maurice Wilks, as chief engineer, while his brother Spencer was managing director. Air Vice Marshal Tedder announced the selection of the Rover company in March 1940 and noted that the Ministry had decided to call in a firm with production experience and suitable plant 'to undertake development manufacture' and to co-operate closely with Power Jets 'to ensure that development designs went along suitable production lines.'

Tedder hoped that there would be 'a very intimate basis of co-operation' between Rover and Power Jets. However, he certainly anticipated the possibility of friction between the firms and in a discussion with Power Jets and MAP officials in April 1940 stated that he required 'complete frankness between the engineers concerned'. He also asserted that some arrangement would have to be made over the rights to future inventions arising from the engine development 'so that there would be no barrier to this.'

By May 1940 difficulties were building up and Tedder noted that he had been worried by the way things were going and that 'Mr Whyte had gone to quite unreasonable lengths in trying to safeguard Power Jets' [commercial] position.' However, 'in spite of much provocation', he intended to keep Power Jets alive and was still resolved not to let it be swallowed by two larger companies. These remarks imply a deep unease with Power Jets on Tedder's part and it is worth noting just how early in the war the relationship between Power Jets and government officials became strained. Whittle also recalled a meeting at this time to discuss patents with Sir Wilfrid Freeman, then Tedder's superior and in overall charge of development and production, who opened the discussion with 'what shall we do about this bloody man Whyte?'

As Rover began to understand the design and gear up for production, their engineering staff naturally began to question design details and to suggest alternative solutions, and this added to the tension in the project. Power Jets considered that Rover design changes were gratuitous. To Whittle's team it appeared that the Rover managing director S.B. Wilks 'was fighting hard to get a commercial position to which he was not entitled' and was 'very persistent' in refusing to agree to safeguards for Power Jets. The Power Jets position was that Rover would make their profit on the production contract and they were increasingly suspicious of any actions which seemed to show that the company was seeking to consolidate commercial rights for the developed engine in post-war markets.

W.E.P. Johnson, the patent agent at Power Jets, and a friend and former RAF colleague of Whittle's, was intensely loyal to the inventor and displayed a deep personal resentment of any action from Rover or from the official side which appeared to prejudice Whittle's rights. His suspicions appear to have had a foundation and, in the early period, it does seem that Rover and other contractors patented some Power Jets ideas. However, more measured responses, even in the face of great provocation, might well have been more productive. By August 1940 Johnson put it to an MAP official that:

The Rover people had not the shadow of an excuse for wanting commercial rights outside the Air Ministry contract ... they were therefore trying to get something to which they were in no sense entitled. I asked him how on earth he expected me to co-operate willingly with people who were trying to get something to which they had no right.

In discussing the relationship between the companies it must also be understood that the Rover company was faced with an enormous task. The enlargement of the Whittle W.1 design meant that the W.2 was a new and unproven design. In spite of Power Jets' jealousy over all the details of the engine this proprietorial attitude seems, in retrospect, hard to justify since the W.2 was not even at prototype stage. G.B.R. Fielden, one of Whittle's most able engineers, suggested many years later that 'the W.2 engine was not a mechanically complete design when the decision was made to put it into production' and that what Rovers got, in the main, was an aerodynamic design which was 'not quite right'. Another Power Jets engineer, Jim Boal, who had a rare continuity of experience of working on the W.2 engine at Power Jets, then at Rover, and finally with Rolls-Royce, also considered that initially 'there was no W.2 design'. Another view from a participant was that 'Rovers were sold a pig in a poke, poor devils. The W.2 had not run [when the decision was made to produce it] … . Absolutely bonkers.'

Thus the situation was that in 1940 the Whittle team contribution to the W.2 was a still-evolving aerodynamic design, which covered the characteristics of the compressor, combustion chambers and the turbine, and a general mechanical scheme for the engine. The actual engineering solutions for the construction of W.2 had not been fully established, but were being developed by extrapolation from the successful W.U. and W.1 and in the light of test results, at both Power Jets and Rovers, with various prototypes of the W.2 design. These revealed a continuous stream of new problems. Rover, as contractor, also had to accommodate input from Gloster engineers on aircraft installation requirements as well as reacting to continuously evolving thinking from Power Jets and from their own engineers on materials, performance improvements and manufacturing solutions. It is not surprising, given the magnitude of the development task, that a sense of ownership should also have arisen from the Rover side. The position over the engine, with regard to production at Rover, was therefore quite different from the well-understood situation where a complex, but developed, product is built under licence by another manufacturer. The Bristol air-cooled piston engines, for example, were manufactured under licence by Gnome-Rhône in France during the inter-war period. In such a case the licensee generally trusts the originator of the design, is most anxious to obtain all the drawings and 'tacit knowledge' which will enable it to succeed, and is, in the early stages at least, most reluctant to deviate from the design and practice of the 'parent'.

However, a hopeful, and peculiarly British, initiative at this time was a unique experiment in inter-firm communication, the Gas Turbine Collaboration Committee (GTCC), which was to involve all the companies that were working on the new engine. The original suggestion was credited to Ernest Hives of Rolls-Royce (who apocryphally remarked that 'as we are giving so much information to the Americans we might as well give it to each other'). However, the main agent in establishing it was Harold Roxbee Cox, and although the normal practice for engine procurement within the Air Ministry and MAP relied on competitive development between firms it was accepted, for the jet, that the problems were so new and so many that a special forum would be helpful. Jack Linnell, as Controller of Research and Development, took the view that it would be worthwhile if it brought 'the various factions within speaking distance of each other'. In fact the committee worked well and Roxbee Cox secured the co-operation of all the firms, arguing that they were ushering in a new age and that it was up to them 'not to fumble the job'. The greatest issue between firms – the question of patenting new ideas – was 'banished from the agenda' on the assumption that it would ultimately be solved, and the committee served as a really valuable agency for the exchange of information on aerodynamic or mechanical problems and their solutions which were new to all the participants. The GTCC certainly made a powerful contribution to the speed with which de Havilland and Rolls-Royce picked up gas turbine work, and to the quality of their engines.

• Chapter 5 •

The 'Straight Through' Engine

In the difficult relationship that existed between Power Jets and Rover, technical disputes could be regarded as almost inevitable, given the magnitude of the task and the fluid nature of the design, although they were perhaps exacerbated by being played out as a kind of three-cornered negotiation between Power Jets, MAP and the Rover company. However, the tension between them increased immeasurably when Power Jets learned in 1942 that the Rover company was attempting a radical redesign of the engine.

The engine, in the form evolved by Power Jets, had an odd feature which independent engineers were bound to question – a double reversal of airflow, with the gas following, in effect, an S-shaped path between compressor and turbine (see Figure 1). The Whittle team had done this to make the engine as short as possible in order to avoid the possibility of the destructive 'whirling' of the shaft coupling the turbine and compressor (swinging out of line like a skipping rope), and to reduce the effect of thermal expansion between the outer sheet metal combustion parts and the inner shaft.

The disadvantage was that the reverse flow arrangement imposed extremely complex shapes for the combustion chambers and associated gas trunking which required highly skilled and laborious sheet metal manufacturing operations. This high requirement for the most skilled type of sheet metal work must have caused serious concern about the possibility of manufacturing the engine in quantity. This is because parts which are machined or cut from solid by lathes or milling machines can be delegated relatively easily to less skilled operatives once the procedures, gauges and tooling have been established. By contrast, forming and welding sheet metal, especially of the heat-resisting semi-stainless grades used in the Whittle engine, is a highly skilled industrial craft which could not, at that time, be mechanised or de-skilled.

Another factor against the reverse flow arrangement was that the two 180-degree changes of direction of airflow were thought to cause internal air resistance and cost a significant amount of power. It appeared early on in W.2 development that the projected power would not easily be achieved and this performance deficit seriously undermined the military case for the aircraft.

The response of the Rover engineers was to rearrange the layout to get rid of the reverse flow feature (see Figure 3 overleaf). The compressor and turbine were kept to the same pattern, but the double-wall

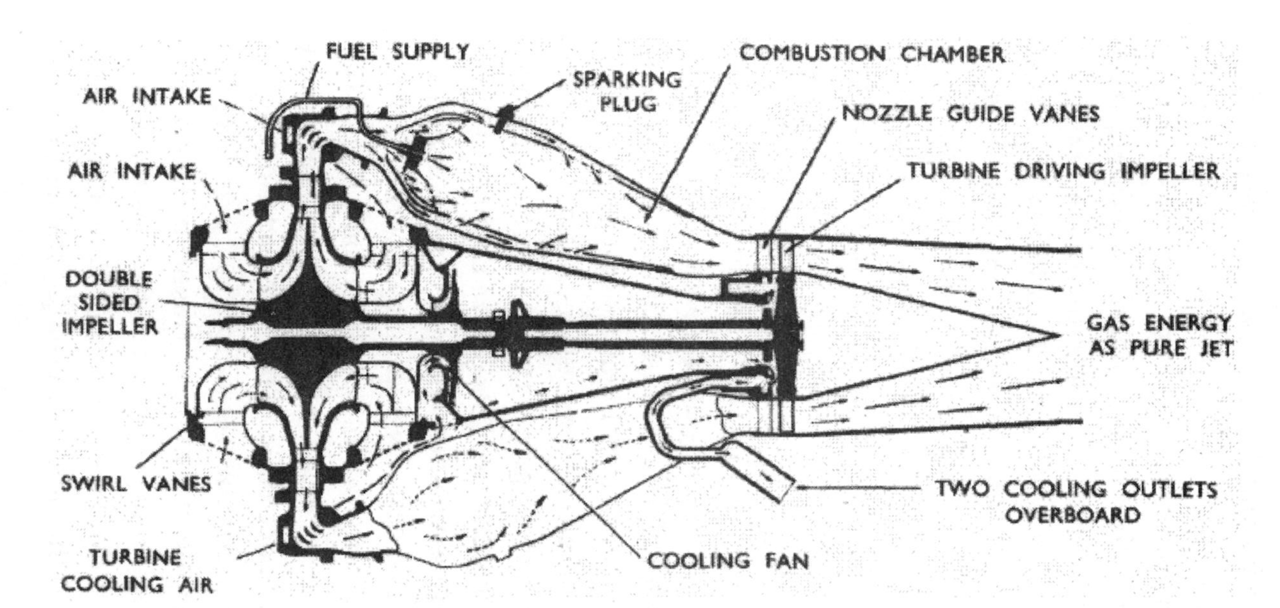

Figure 3: A 'straight through' centrifugal compressor engine. The double-sided compressor (to left) and turbine (right) are virtually identical to the Whittle design, but the shaft joining them is longer. This is a later Rolls-Royce Derwent engine, but the basic architecture is similar to the Rover design.

counter-flow combustion chambers were replaced by 'straight through' cans which had the form of simple cones and were far easier to make. The only additional problem brought by this rearrangement was the requirement to provide a third bearing to support the longer shaft in the middle, and a sliding coupling to allow the shaft to adapt to the thermal expansion of the hotter outer engine parts. Both of these considerations were accommodated by an ingenious coupling devised by the Rover engineer supervising the project, Adrian Lombard, although Whittle later made a veiled suggestion of plagiarism by noting that 'for some time before and during the controversy, I had standing on my desk a wooden model of a shaft coupling which we were proposing to use in our own "straight through" engine when we could get around to it.'

It now appears that Rover had, in this period, encouragement for the redesign from Bulman (now Director of Engine Development and Production at the MAP) and from his deputy, Major A.A. Ross. Certainly since mid-1941 Bulman and Ross were beginning to question Power Jets' ability to get the engine right, as seen from a meeting of the Gas Turbine Collaboration Committee at which Ross remarked that 'in the development of the W.2B engine Wing Commander Whittle has been regarded as qualified to speak "ex cathedra". Any suggestion that the Rover Co. should be given licence to embody their own views even in details of a purely mechanical nature has been accepted without enthusiasm. It is now clear that Wing Commander Whittle himself is far less assured of his own aerodynamic features and is therefore anxious to obtain help from anyone and is looking especially to Rolls-Royce.'

The design work and construction of this Rover straight through engine was carried out in secrecy, and news of the development was kept for some time from Power Jets and from the more junior MAP representatives at the firm. This was because the Power Jets view that MAP were unfairly partial to Rover was also matched by a feeling at Rover that sections of the MAP were overly sympathetic to Power Jets and that great pressure would be brought to bear to make them desist from the straight through arrangement which they genuinely felt to be superior.

The Rover company first officially presented the design (known as the B.26) at a meeting on Rover premises in February 1942 when it was described as a 're-arranged assembly for ease of production' and, to the subsequent fury of Power Jets personnel, as 'the first serious attempt to productionise the design'.

The MAP minutes of the discussion recorded the Rover view that 'until an attempt had been made to manufacture the W.2.B. on mass production lines it had not been possible fully to appreciate the inherent difficulties, but the need for a revision of the assembly features was daily becoming more and more apparent.' Among a (large) number of advantages for the new arrangement was the fact that the impeller and turbine assembly could be assembled and balanced as one unit and then mounted in the engine ('out of balance' forces could easily destroy an engine at the new high rotational speeds the jet reached). In the Whittle design the rotating parts were balanced in a test rig, then stripped, and laboriously reassembled in the actual engine, with every component put back in the balanced position and every turbine blade marked and numbered to make sure it was returned to the correct sector on the wheel. The sheet metal work for compressor air ducting and the combustion chambers, as already discussed, would also be far simpler, and the number of nuts and bolts (900 in the Whittle version) which held all these together would be considerably reduced.

Whittle took the view that a 'straight through' engine was under consideration by Power Jets anyway, but that time could not be spared from development of the proven layout in order that jet fighters should reach the RAF as early as possible, and that, in any event, Rover had no remit to do development work. The redesign seemed to bear out all their suspicions about Rover's intentions 'to gain a commercial position to which they are not entitled' and in April 1942 W.E.P. Johnson set out Power Jets' many grievances in a long memorandum for Dudley Williams and the Power Jets board entitled 'The Rover Independent Development'. Apart from copious criticisms of Rover from a technical point of view ('The technical staff has never found it easy ... to collaborate with Rover. This word "collaboration" has always meant, in practice, to carry, and to carry uphill at that'), the paper certainly made explicit Johnson's own views about the bad faith of MAP officials and is a clear symptom of the 'general xenophobic feeling' referred to earlier – a feeling which spilled out beyond these internal papers into the general communications of the company with MAP. A relatively small proportion of the paper shows its general tone:

> *The pith and the marrow is this: the National effort is being impeded by the actions of Rover and the Official condonation thereof ... .*
>
> *It seems to be fairly clear that in the middle of February this [Rover straight through] engine must have been well on the way to completion and it is inconceivable that Major Ross was not aware of the fact. If he was aware of it, then he has shown a measure of duplicity not only in his attitude at the meeting itself, but also in his Minutes which seems to be so incredible in an experienced*

Government Servant that I hesitate to express it, let alone accept it … . The real impropriety is an ethical one and the Department [the MAP] is the guilty party, as I see it … . Broadly and nationally, these inexperienced and inexpert people should not be wasting their capacity on such matters … .

I believe the incident to demonstrate clearly that there is collusion between a Dept. and Rover, to defeat the objects of Collaboration, a secondary effect of which will be to enable Rover to build up prestige and weight as designers and originators, such as will stand them in good stead after the War. There can in my opinion be no doubt whatever as to the collusion.

Johnson proposed, among other 'reforms' to be requested from the Ministry, a new policy 'that Rovers act as producers only, on a Shadow basis'. This showed a certain lack of realism. 'Shadow' manufacture, as we saw when discussing pre-war RAF expansion, relied on taking a highly developed aero engine, such as the 14-cylinder Bristol Hercules, and replicating all the established manufacturing, gauging, fitting and assembly procedures in a new plant with new personnel. In itself this was a highly demanding exercise, but it relied crucially on the parent firm having finalised the design. Under its great design autocrat, Roy Fedden, Bristols were meticulous in detailing their processes, materials and know-how in a 'Bristol Engine Bible'. The situation with the Power Jets/Rover W.2B engine was not remotely like this. Thrust and reliability were still far from the required standard and a formidable task of development engineering still lay ahead.

Whittle and his team made vehement representations to MAP about the Rover straight through engine and Whittle explicitly took up with MAP the direct attack on the judgement of government servants that Johnson was suggesting in internal company papers. He asked for a full enquiry to examine, among other things, 'the qualifications, judgement, and experience, of the Civil Servants who have had such a powerful influence on the project', offering to step down from the jet project 'if my judgement and experience in such matters are believed to be inept'. The eventual outcome of the dispute was that Rover was allowed to continue with the straight through engine 'but that this should not be at the expense of the W.2B production programme'. Bulman, for his part, later said: 'We had all long since realised that Whittle was his own worst enemy, quick to invest every discussion with the venom of suspicion, scavenging through letters and minutes of meetings for odd words or phrases which he could pick on to suggest that they were deliberately ambiguous and revealing of a sinister influence behind the scenes, determined to "do him down" lest his jet become damaging to the piston aero engine.'

Whatever the merits of Power Jets' moral case, the Rover B.26 was a rational solution and, moreover, it became the eventual successor to the Whittle layout. When Rolls-Royce eventually took over the Rover production plant (of which more below) they manufactured only enough reverse flow Whittle W.2Bs (under the name 'Welland') to equip twenty Gloster Meteor Mk 1 aircraft before converting to the more aerodynamically efficient straight through engine, based on Adrian Lombard's Rover B.26, and called the RB 37 Derwent I in Rolls-Royce terminology.

Although Power Jets complained of the 'inexpert, inexperienced' Rover people being allowed design responsibility although 'not competent to do the work' we should note that Lombard remained the supervising engineer on the project with Rolls-Royce so there was absolute continuity from the Rover straight through proposal to the highly successful Rolls-Royce Derwent (of which 500 were made) which raised the speed of the Meteor from 415 to 470 mph. Jim Boal also recalled the Rover design team, which included some powerful analytical talent with J.S. Jones, a mathematics Senior Wrangler from Cambridge, as very good, 'and not given credit'. Adrian Lombard went on to become the most eminent director of engineering at Rolls-Royce in the post-war period when the company's reputation in gas turbines was firmly established with engines like the Avon, Spey and Conway.

There is an almost tragic quality of hubris about this episode, epitomised by Johnson's conclusions and policy recommendations, particularly that 'it is our duty to badger the MAP, at the risk of unpopularity, into seeing the way it is heading.' (Emphasis added.) He suggested:

> *I am certain that if it is possible to put the whole position clearly to responsible and impartial officials, a radical revision of affairs must follow. We should, however, make it clear that our feelings are as much against certain Officials as against Rover, for these parties are, on the evidence, indistinguishable.*

The episode highlights what can only be seen as an extraordinary sense of self-importance at Power Jets, and of naivety, in imagining that they, as a relatively tiny state-supported company, could force the removal or possibly the disciplining of trusted MAP officials who were successfully running huge production programmes with the mainstream engine producers. Power Jets were also unable to see that these officials, whose judgement they doubted, were equally motivated and highly committed. Air Marshal 'Black Jack' Linnell, Controller of Research and Development at MAP, and Bulman's boss, was considered 'a tiger to work for' who drove himself, and his staff, mercilessly, and would stay up 'literally all night' reading technical reports on aircraft in development. To keep up this pace, it was rumoured he resorted to 'Tail-end Charlie' pills – Benzedrine, a form of amphetamine quite frequently given, in the Second World War, to combat personnel and rear gunners on bombing operations to promote wakefulness and alertness.

By a strange coincidence, Whittle too had become a Benzedrine user, apparently addicted inadvertently through frequent use of inhalers which were then sold as an over-the-counter remedy for nasal congestion. Side effects of the drug are now considered to include feelings of depression, tension and even paranoia. The amphetamine issue may be of minor significance but Whittle did recall feelings of 'tension, anxiety and insomnia' and Golley recorded 'a downward path with sleeping pills, tranquillisers and stimulants'.

Whittle also suffered terribly from boils and from persistent eczema and inflammation of his ears – all presumably the result of tremendous emotional stress. Indeed, there seems little doubt that the emotional temperature of the whole project at Power Jets was extraordinarily high and rather different, for example, from the milieu at Rolls-Royce, where technology was being pushed just as far and as fast, but without the same sense of embattlement. Thus L.L. Whyte, who was a real asset, and the only member of the team with the stature and managerial expertise to hold his own with MAP officials, fell out with Whittle and resigned as Chairman and Managing Director of Power Jets in July 1941. He recorded merely that 'difficulties had accumulated … . Early in that month my association with Whittle, which had lasted nearly six years, came to an end. There were too many difficulties between us, and I was not sorry to leave Power Jets.'

In this period Whittle also fell out with Isaac Lubbock, of the Shell combustion laboratories in Fulham, who had developed the combustion chamber which solved the problems that had been encountered up to then in getting stable combustion, allegedly because Lubbock patented the 'Shell' combustion chamber for his company. Although Whittle, in Jet, described this merely as 'a combustion chamber of approximately the same size and form as that used in the engine', Lubbock's son was adamant that the work was specifically for Whittle, and that Lubbock had been 'a father figure' until this rift, amplified apparently by L.L. Whyte (described as 'pure poison' by Lubbock's assistant) who threatened Lubbock with prosecution under the Official Secrets Act for seeking the patent. Nevertheless, Whyte was a loss, for in spite of earlier friction with officials he had the 'savoir faire' to represent Power Jets to MAP and to Rover. In his place Williams and Tinling became joint managing directors although, in Fielden's view, they were 'worthy, straightforward but not highly intellectual', and Whittle had 'no one to lean on ... if only Power Jets had had a commercial managerial man.'[3]

Williams, Tinling and Johnson, while intensely loyal to him, echoed and amplified Whittle's own suspicion and jealousy about outside intervention, intellectual ownership, modifications by Rover and so on, rather than moderating his reaction. Their continual interventions with MAP after Whyte had gone were almost certainly unhelpful to Whittle's cause. Official and company papers suggest that the management came to be seen as suspicious, resentful and 'prickly', and became increasingly unpopular with officials. Whittle gives suggestive details of a meeting with Sir Wilfrid Freeman on 11 December 1942, who had by then returned to the MAP as Chief Executive, at which Freeman 'seemed to be antagonistic to the Board of Directors', suggested that Williams and Tinling had ceased to have any useful function, and 'went so far as to say he would have them "called up" [for military service].' Johnson too was mentioned in connection with call-up.

The senior management of the company was not a good psychological mix, and it also held an exaggerated view of the company's potential and bargaining strength. Certainly there was little awareness of the danger they faced, for even as these controversies with MAP were running, Ralph Dudley Williams wrote to Maurice Bonham Carter looking forward to the time when 'the Management would be able to say we are practically inheriting the Rolls-Royce position as the leading aero engine manufacturers of the day.'

Because Power Jets had not evolved over years in contact with government departments it had not developed the mechanisms, habits and experience for dealing with them and, from the point of view of some officials, Power Jets was a problem. Many public servants approached the venture in an open-minded spirit, accepting that a new engine must be the product of unusual minds. However, others were concerned by the continually deferred W.2 programme, the poor relationship between Power Jets and the Rover company and the combative approach of Power Jets towards MAP.

There was, after all, the example of the de Havilland jet engine – promising, powerful and actually overtaking the stalled Whittle programme so that de Havilland engines replaced Power Jets ones for the first flight of the Gloster Meteor prototype on 5 March 1943. The de Havilland engine department, led by the freelance engine designer Frank Halford, had started to study the gas turbine in early 1941 at the request of MAP when, according to Bulman, Tizard, 'harassed with the vicissitudes of Whittle, cast around for some more normal character to tackle the design of a jet engine.' The de Havilland team aimed at a larger, higher-thrust engine, also of 'straight through' design, suitable for a projected single-engined fighter (the D.H. 100, which became the de Havilland Vampire). They had the benefit of access to Power Jets' data and experience and membership of the Gas Turbine Collaboration Committee to keep abreast of all jet work in other firms, while the aerodynamic design for the engine was developed at RAE Farnborough, in the light of experience with Power Jets designs. Nevertheless, the de Havilland team did a superb job on the engineering and mechanical design, building a functioning engine which demonstrated 3,000 lb of thrust in June 1942 – less than two years after work started. At this time the Rover-built engines were restricted to 1,000 lb thrust and, additionally, on account of impeller bursting problems, were restricted to taxying trials only. De Havilland certainly had a 'flying start', but Power Jets itself had received a huge amount of assistance from the RAE, which had seconded some of its best compressor and turbine specialists to the project and subordinated work on its own axial flow F2 engine.

Although the straight through engine was put to one side by Rover following the fracas between Power Jets and MAP, a stream of difficulties and complaints continued to emanate from Power Jets.

Moreover, the results obtained with the basic W.2B over the next ten months were poor. Against this background of disappointing engineering development Power Jets kept up a continual campaign of argument with MAP. Much of this was concerned with the division of design responsibility between Power Jets and Rover and with Power Jets' irresolvable claim for a formula which would recognise them as the ultimate design authority. This was problematic, since the design was continually evolving, and since creative engineers (as Lombard and his Rover colleagues clearly were) were bound to have personal or original views on possible solutions.

MAP attempted to resolve these disputes by drafting 'terms of reference' which were intended to regulate affairs between Rover and Power Jets. The growing exasperation of Air Vice Marshal John Linnell, now the Controller of Research and Development (CRD) at MAP, and in overall control of the jet engine programme, can be glimpsed from his letter to Power Jets enclosing these terms of reference and making the appeal that:

> *It is quite impossible to define with complete precision all the points which may arise. The Terms of Reference ... must ... be used as a general guide and interpreted in a liberal sense on matters which are not explicitly covered ... A similar letter has been sent to Mr S.B. Wilks of Rovers.*

Linnell observed that the urgent need to produce a jet aircraft had 'forced us into a position where we must attempt to manufacture the W.2B concurrently with the development and testing of experimental models', but reflected that this process 'lends itself to misunderstandings as regards responsibilities' and that success would depend on 'the fullest possible collaboration'. He also announced the formation of a Technical Committee set up by MAP, comprising an MAP chairman, representatives of the firms and the Resident Technical Officers (Ministry men who 'lived with' Rovers and Power Jets). It would rule on all technical matters, modifications and points at issue between the firms, including the 're-arranged layout of W2B known as [the straight through] B.26', and 'authorise the embodiment of modifications on the production line'. Williams protested that this meant 'there was someone sitting above who could rule on all our work on an imperfect understanding of it', and wrote to Bulman a long historical summary of past grievances. Bulman replied:

> *Thank you for your letter of July 23rd ... . Fortunately my many other correspondents do not regale me with such lengthy epistles, though their matter is frequently of equal moment. Following our frank and to me very happy and most useful talk on July 20th, I think no further purpose would be served*

in further inquest on the past, with so much for all of us to do now .... I attach a copy of the working procedure for the MAP Technical Committee which has been approved by CRD who is not prepared to have any further discussion either on the terms of reference or on this particular document.

Williams continued to object that the proposed procedure 'was a great disappointment to us' and stubbornly suggested that the new procedure meant that Power Jets must 'submit every idea for improvement' to Ministry authority and 'may not proceed to test without permission'. 'I fear that "row" you asked for when we last met is not long in coming. May we see you soon and get it over?' Bulman replied, one senses, wearily that the policy:

... does not introduce any alteration in the existing procedure ... in the evolution of new ideas and/or the trial of experimental features in your engines in your works. The document is clearly intended primarily to clarify the situation affecting our mutual relations with the Rover Co, but not to modify them ... .

The Terms of Reference ... have been laid down by CRD after immense expenditure of time and discussion ... there is nothing I can see to be usefully added ... in relation to the actual progress of development and production of the W.2B engine, which is our sole objective, and not argumentation about the past, of which there has been more than enough already.

We have seen that there were factions within MAP with respect to Power Jets, although the company's actions progressively distanced its supporters. Whittle's admitted brilliance had won support from Tedder (before he left to become deputy commander of the RAF in the Middle East from December 1940), from Tizard, Pye and, perhaps, from Linnell. Bulman, though, as the most important commissioning agent for RAF engines in the inter-war period, was a strong supporter of the established companies, particularly Rolls-Royce and Bristol, and was quite sceptical about Power Jets. This polarisation also reflected a division of responsibility within MAP for the jet. Bulman was responsible for engines actually commissioned for RAF service while, as we have seen, Power Jets came under the immediate direction of the DSR (Director of Scientific Research), through Harold Roxbee Cox – a situation that reflected the experimental nature of the jet and the original sponsorship of it by the Engine Sub-Committee of the ARC and by the DSR. Nevertheless, the firm had frequent contact with Bulman as design originator for the engine that Rover was building for service, and this dual-track approach from MAP must have added to the pressure on Whittle and Power Jets. But by late 1942 the stock of goodwill within MAP towards the firm was largely exhausted.

Sometime earlier Whittle had succumbed to the first of two nervous breakdowns. Sir Rolf Dudley-Williams recalled what was, presumably, the start of this: when a colleague, L.J. Cheshire, told Whittle that a British warship had been sunk, Whittle threw a chair at him, though another participant recalled that all the staff were intensely loyal to Whittle and protective of him 'when he was peculiar'.

This presumably was the same occasion recorded by Bulman, when news came in that Japanese forces had sunk the battleship Prince of Wales and the battle cruiser Repulse. According to Bulman (perhaps not an impartial narrator), 'this had the effect of making poor Whittle go completely out of control, screaming, "Well don't blame me. I couldn't help it." It was the culmination of increasing nervous tension on an inherently temperamental personality. He was attended by the senior RAF specialist in neurology and was away for a month, returning to duty too soon.'

Kings Norton recalled visiting Whittle in hospital: Whittle, pointing at his head, said, 'They are giving me electric shocks.' Kings Norton went to the chief medical officer of the RAF to have the treatment stopped, encountering the argument that it was needed to get Whittle back to normal. According to his own recollection, Kings Norton replied, 'He never was normal.' This would have been a very early use of electroconvulsive therapy (first tried in Rome in 1938), and is evidence of the urgent desire in the RAF for the jet engine to succeed. It must have been virtually experimental and was certainly completely inappropriate for someone whose major problem seems to have been acute stress. However unfair it may seem, some officials must certainly have considered his nervous troubles as evidence of unfitness to control the jet programme and a participant close to the programme later suggested that 'most of the officials regarded Whittle virtually as a nutcase'. L.L. Whyte described the episode more poetically, recalling:

> *Unbroken tension and excitement, with its result in nerves and illness. At one critical stage Whittle himself could stand it no longer and as he lay in bed day after day the whole enterprise was shaken by the appalling doubt: had too much been gambled on one man and had that man taken on too much?*

The disillusionment with the jet programme is reflected in Linnell's note in November 1942 to Sir Henry Tizard, in which he expressed disappointment with jet progress and anticipated that the F.9/40 (Gloster Meteor) aircraft fitted with W.2B power units, then achieving 1,450 lb thrust, would be 'of such inferior performance in climb ... and of so little superiority in speed' as to be unacceptable and 'useless for operations by the time it is introduced.'

Tizard agreed that the jet position was disappointing but said that although 'the gamble of preparing for the production of the W.2B engine on a large scale has not quite come off' he felt that 'even with all the mistakes, human and otherwise' it had been justified. Shortly afterwards Linnell summarised the problems that had befallen the jet project for Air Marshal Sir Wilfrid Freeman, who had returned from serving as Vice Chief of Air Staff at the Air Ministry to become Chief Executive at the MAP. Linnell observed that 'bitter experience' with the piston-engined Typhoon aircraft (designed by Hawkers but produced at Glosters) and the W.2B had 'proved that the only sound way to introduce a new design is for the design firm to be charged with the initial production.' The major causes of the troubles with the W.2B he attributed to this mistake, to the attempt to produce a completely undeveloped design and to 'undue optimism'.

Linnell recommended that Rolls-Royce should take over Power Jets with their factory at Whetstone and run it as the jet section of Rolls-Royce, while Rovers should 'finish out the salvage of the W.2B engine from which I should hope to build between 150–200 training and development engines.'

Freeman then advised Cripps that he had ordered the production of W.2B engines to be stopped at 200 examples and that the Meteor aircraft run would be cut short at about 50. 'Fifty [aircraft] should be about right', he noted, in a comment which certainly shows how little faith there was in potential reliability, 'as I imagine we will run through the engines pretty quickly.'

Sir Wilfrid Freeman then proposed to Whittle a take-over of Power Jets by Rolls-Royce in December 1942 but seemed unwilling, or unable, to force it through. The reason for this is not entirely clear but the scheme probably foundered on Whittle's claim to be Chief Engineer of a future integrated scheme and Freeman's view that 'he could not very well put Rolls-Royce under [Whittle's] orders'. Whittle recollected from this meeting 'a further harangue, in which he still had not explicitly stated that Power Jets were to be handed over to Rolls-Royce, though he had again inferred it.' Perhaps significant was the absence of any administrative or financial mechanism open to MAP. It could nationalise underperforming companies (Cripps nationalised the Short aircraft company in 1943 for poor production performance) but it had no remit to force the acquisition of one private company by another. Instead, Rolls-Royce took over the Rover production facility at Barnoldswick from early 1943, leaving Power Jets as an independent concern to carry on

1. Frank Whittle photographed in his study in 1946. The models are the Gloster-Whittle E.28/39 aircraft, the Gloster Meteor and a Power Jets W.2/700 engine. (Science Museum/Science and Society Picture Library)

2. The first ground test engine (the W.U.) at Power Jets' works in Lutterworth, 1938. A second-hand light car engine was used as a starter motor. Today, the engine is in the Science Museum, London. (Science Museum/Science and Society Picture Library)

3. Above: rotor of Power Jets W.2/700 showing the double-sided centrifugal compressor (left) driven by the turbine wheel. For comparison, below, the rotor of an axial flow jet engine (the Metrovick F2). The aerodynamic design for the F2 compressor was done at the Royal Aircraft Establishment, Farnborough, under the leadership of Hayne Constant. (Science Museum/Science and Society Picture Library)

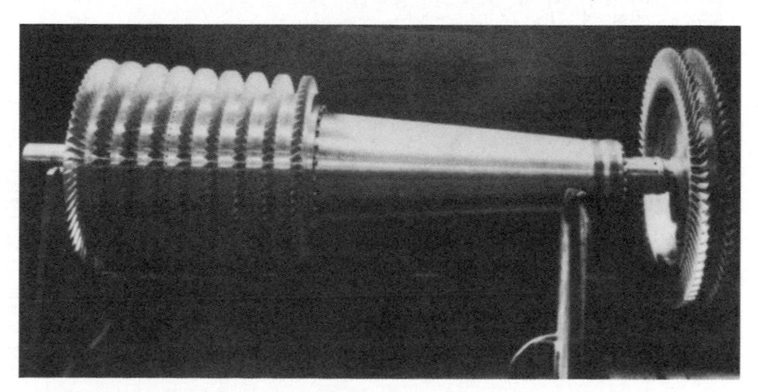

4. Building the myth. After the war, a scene showing Whittle and colleagues testing the first engine was recreated for the official film *Jet Propulsion* using the old testbed. (Science Museum/Science and Society Picture Library)

5. The Gloster-Whittle E.28/39. It lifted off briefly while doing taxying trials in April 1941 and made its first true flight of 17 minutes at Cranwell in May. It had a long research career during the Second World War flying with several different jet engines. Today, it is in the Science Museum, London, alongside the W.1 engine which powered it for the first jet flight in Britain. (Science Museum/Science and Society Picture Library)

Jet-propelled airplanes are going into mass production

BRITAIN HAS FIGHTER WITH NO PROPELLER

All tests passed, speed is colossal

ENGLISHMAN INVENTED IT

Express Air Reporter BASIL CARDEW

THE Allies now have a fighter aircraft that can fly at tremendous speeds—without a propeller. It is a jet-propelled machine, something entirely new in practical aviation.

The news was broken last night in a joint statement by the R.A.F. and the U.S. Army Air Force. They announced that the machine is about to go into mass production for both forces.

When this fighter goes into action its enormous speed will probably call for a complete change in the known tactics of air warfare.

The genius of the plane is in the power unit. The system of jet propulsion is roughly this: air is sucked in through a scoop in the nose, compressed, heated and shot out behind at high velocity.

The jet forces the aircraft forward at a velocity which may not yet be disclosed.

This fighter, invented by an Englishman, has taken ten years to produce. The experiments, begun in Britain, were carried on in the United States.

Began 10 years ago

Work began on the engines in 1933 by Group-Captain Frank Whittle, the inventor, a pre-war R.A.F. man.

His first jet-propelled engine ran successfully after four years' research work, in 1937. The Air Ministry at once saw

'Like something from Mars'

WHITTLE WORKED ON IDEA 14 YEARS

Express Staff Reporter

GROUP-CAPTAIN FRANK WHITTLE was 22 years old when the idea of jet-propulsion became for him the most important thing there was.

That was 14 years ago. Fourteen years of drawing boards and theories, experiments and tests, set-backs and success.

Now Frank Whittle, 36 years old, married, with a son aged 12 and another aged nine, lives in the Midlands...

THE INVENTOR

GROUP-CAPTAIN FRANK WHITTLE

THE TEST PILOT

OVER 500 m.p.h. POSSIBLE

AT least three of the warring nations, Britain, Italy and Germany, have been searching for a successful jet plane for many years.

THE JET-

FROM PAGE ONE

FINEST TEST PILOT

7.1.

POWER JE

PRESS CU

1944

6. The headline story from the *Daily Express* for Friday 7 January 1944. (Express Newspapers Ltd)

7. The de Havilland Comet 1 prototype in BOAC livery, probably some time in 1950. After public demonstration at the Farnborough Air Show in September 1949, it put on record high speed proving flights at over 420 mph to Rome, Copenhagen and Cairo. (Science Museum/Science and Society Picture Library)

8. Vickers VC 10 airliner used for flight tests of the new Rolls-Royce RB 211. The substitution of the two Rolls-Royce Conway engines on the port side by a single RB 211 underscored the step-change in size and complexity that the new engine represented. (Rolls-Royce Heritage Trust)

research and development of continuation engines to the basic Whittle design.

Bringing Rolls-Royce into the project quickly transformed the outlook for the Whittle engine. As Chief Engineer for the jet team Rolls-Royce appointed Stanley Hooker, their engineering scientist who had done much to keep the power of the Merlin piston engine improving throughout the war (from just under 1,000 bhp to over 2,000 bhp by the end) by steady improvements to supercharging.

Ernest Hives, at the head of the company, nurtured a company culture with an exceptional attitude to engineering quality and a ruthless approach to research and development. Kings Norton remarked that 'Hives believed in Rolls-Royce the way some people believe in God.' His tough pragmatic view is captured in his letter to Linnell in March 1943 which promised to treat the jet 'as just another aero engine' and not as a piece of scientific apparatus. 'We do not look upon the turbine engine as a new secret weapon, it is just another way of pushing an aeroplane along, except that at the present time it is not as good as the conventional engine.' He also noted for their next meeting that:

> *I shall be bringing with me a report which will give the performance of the F.9/40 [Meteor aircraft] with various stages of progress [of the jet engine] and the dates at which we think these will be*

available. We want it to be understood that these are Rolls estimates, and therefore have to be taken seriously … . We are satisfied that we are going to make a success of it … the progress will be in keeping with our reputation.

Rolls-Royce did in fact develop the W.2B to become a reliable unit but, as we have seen, it was quickly replaced by a development of the Rover-originated straight through variant.

• Chapter 6 •

The Origins of the Jet Engine Programme in the USA

In the USA, as in Britain, the origins of the gas turbine engine were much more complicated than has generally been appreciated. Nevertheless, also as in Britain, the Whittle engine played a crucial role in stimulating development.

While still a student at the University of California in 1895, the engineer Sanford Moss had conceived a gas turbine, intended as a stationary unit for industry, commenting later wryly that 'like most of the other inventors, [I] ... at first thought [I] was alone in the gas turbine field'. Moss developed the gas turbine to the hardware stage at General Electric (GE) in Lynn, Massachusetts, in 1904, but concluded that 'as the power for compression was more than the turbine power ... the experiment was a flat failure'. However, Moss went on to become a central figure in the development of the turbo-supercharger for piston aero engines by GE at Lynn. This work, intriguingly, became an important link in the transfer of Whittle's ideas to the USA.

During the 1930s various turbine schemes were considered in America. Most notable were those at

Northrop for a complex turboprop (the Turbodyne) and Lockheed's interesting jet project, the L-1000, conceived by Nathan C. Price, formerly a steam turbine engineer, who had worked for a while with the Doble steam car concern, which was aiming at a 600 mph aircraft flying at 50,000 feet. However, none of these ambitious schemes could be funded by the inventors or manufacturers, and the US army and navy were not tempted, before the war, to back a new type of engine and underwrite the research and development. As discussed in Chapter 2, the decision in Britain to back the powerful but thirsty Whittle jet was propelled in part by Britain's vulnerability to air attack from Europe and the short warning time of the approach of hostile bombers. This was not a strategic concern in the USA at the time. It was the impending entry of the USA into the war, and the discovery that active, well-advanced jet work was going on in Britain and in Germany, which acted as the spur.

The actual circumstances in which service and industrial people in the USA became aware of the British work are still somewhat mysterious. For example, it has often been suggested that when General Henry 'Hap' Arnold, as head of the US Army Air Corps (USAAC),[4] visited Britain in the spring of 1941, he was told about the Whittle engine by officials and, it is said, 'was astonished when, by chance', he learned that British jet engines were in an advanced state of development and were 'about to be flown'. It has also been suggested that he witnessed taxi trials of the E.28/ 39, and, on his return, asked for a competent US engineer to be sent over to study the Whittle engine. However, the coming of Lend-Lease, and the supply to Britain of turbosupercharged American aircraft for use by the RAF, had already brought USAAC and GE staff over to Britain for engineering support before Arnold's visit, and the choice of an American engineer to investigate the jet fell, it is said, on 'the ideal man', GE engineer D. Roy Shoults, then in Britain as GE's technical service representative for the turbosupercharger.

Alternatively, according to Schlaifer's generally authoritative account, D. Roy Shoults 'picked up from various sources enough information to conclude that turbojet engines were being developed' before Arnold's visit and passed this knowledge to Colonel A.J. Lyon, technical representative of the USAAC in Britain. Shortly after this, the MAP showed these two the whole British turbojet programme. According to this version, this information 'gathered on the ground' in the UK was then passed back to senior figures in the USAAC and to Arnold, prompting his interest and enquiries in Britain.

In spite of the attention given to Arnold's contact with the Whittle work, it should be noted that he had, apparently, already been spurred by generic intelligence reports of jet and rocket work in Germany and, before his visit to Britain, had asked the Chairman of the National Advisory Committee for Aeronautics (NACA), Vannevar Bush, to set up a 'Special Committee on Jet Propulsion' in February 1941. This was placed under the direction of retired former NACA Chairman William Durand, who proved 'an exceptionally energetic chairman'. By July 1941, the committee was sponsoring three projects: a pure jet engine from Westinghouse, a ducted fan/jet from Allis-Chalmers, and a turboprop, the TG-100, from the GE plant at Schenectady. All these projects used axial flow compressors, probably because intelligence now was indicating that this was the German approach. The projects were also independent of Whittle's work at Power Jets and in fact some teams, such as that at Westinghouse, appear to have been later deliberately kept in ignorance of it in order to foster more than one line of development.

However, there is another twist to this story which Edward W. Constant has drawn out, and it is surprising that its implications have not been generally appreciated. This is that, through their industrial links, GE were well aware of work on the Whittle jet that was going on at the British Thomson-Houston factory in Rugby, England, well before 'Hap' Arnold or indeed Roy Shoults were made aware of the work at Power Jets. The links between GE and BTH, as manufacturers of steam turbine generating sets, were close; in fact British Thomson-Houston had been formed in 1894 to exploit the patents of the American Thomson-Houston company (an antecedent of GE). By this stage BTH were becoming quite possessive about the test engine they were building for Whittle (describing it in literature as the British Thomson-Houston turbojet engine) and they showed it off to visiting GE engineers in 1939.

As a result of the information they passed to him on their return, Sanford Moss wrote in September 1939 to fellow engineer Dale Streid at GE's Schenectady plant that:

> *Mr A.R. Smith has just returned from England and he and his assistant … saw an exhaust gas turbine at the BTH factory for airplane service … . We have the idea that the apparatus involved a centrifugal compressor with ramming intake driven by a gas turbine wheel which furnishes a jet for jet propulsion … .*

This was a perfect description of the Whittle jet, although Moss also speculated as to whether part of the combustion gas bypassed the turbine to produce the jet directly. Moss suggested, 'we have in mind going to Wright Field [the test establishment for the USAAC] with a similar proposition and want to know all we can about the British outfit before we start ... [We] would like to present some sort of proposition to Wright Field now.'

It appears that no proposal was actually put to air force staff but, perhaps as a result, GE's Dale Streid shortly afterwards did a theoretical study on the possibility of jet propulsion for the then very high speed of 450 mph. Another intriguing point raised by Constant is that earlier, the Lynn group frequently tested turbochargers by inserting a combustion chamber between the compressor and turbine and so had been 'inadvertently running a turbojet'. He notes that in spite of Moss's continuing sense that the gas turbine was still premature, based on his own earlier disappointments, 'some of the younger engineers wanted to build a true turbojet'.

The GE advanced engineering staff were in frequent and close contact with air force technical staff from Wright Field over the development and operation of turbochargers, and the clear implication of these ideas and contacts, even though no turbojet programme at first resulted, must be that both GE and the US Air Force were increasingly aware that a gas turbine jet engine would soon be feasible. This reading of events also suggests that when both GE and US Air Force personnel reached Britain during the Second World War, they already knew much more about work on the Whittle idea, and the general feasibility of the jet engine, than has generally been admitted. The confusion about the sequence of events perhaps results from this and a certain coyness, in American circles, to admit how much was known about the secret work of an ally.

In spite of the various gas turbine leads in the USA which have been discussed, knowledge of the Whittle programme suggested that the quickest way into turbojet development was to build on the British experience. Arnold asked for the creation of a working group to facilitate transfer of the British knowledge to the USA in the mutual wartime interest of both sides. The British representation was at a high level, including Air Marshal Linnell at the Ministry of Aircraft Production and Harold Roxbee Cox who was directing the work at Power Jets for the MAP. General Arnold, it is said, selected GE to receive this knowledge, partly because the established aero engine firms of Pratt & Whitney and Curtiss-Wright were fully committed to producing existing types of piston engine but crucially because of its experience with the turbosupercharger. A turbosupercharger (today usually called a turbocharger) consists of a turbine, driven by the exhaust of a conventional piston engine, coupled to a centrifugal compressor which is used to pressurise the inlet air for the piston engine. Ideally, this arrangement exploits and recovers 'waste' energy in the piston engine exhaust, to maintain altitude performance to the piston engine which loses power as it ascends into thinner air.

The turbosupercharger had become very much a GE speciality, and so pre-war aero engine development in the US differed from Britain, where mechanically driven superchargers were standard. Turbosupercharger experiments had been tried at the RAE in the 1920s but had not been followed through. This was, in part, because finding alloys for the red-hot exhaust turbine blades seemed an awesome task, although in the USA it had been tackled using new alloys with high nickel, chromium and molybdenum content such as 'Stellite' primarily developed for piston engine exhaust valves.

On 1 October 1941, a B.24 Liberator bomber crossed the Atlantic bringing to the GE plant at Lynn a set of drawings for the Power Jets W.2B engine, which the Rover company was then endeavouring to build, as well as an actual engine, the W.1X test unit. This was effectively a duplicate of the W.1 engine, used for the first flight of the Gloster E.28/39, built with a spare rotor for the W.1 and non-airworthy parts which had been rejected for the W.1. Nevertheless, it had proved invaluable for ground test work. (It still survives and today can be seen in the Jet Aviation gallery at the National Air and Space Museum, Washington.)

It should be noted that the transfer of Power Jets design and experience to the USA was not simply a question of shipping over drawings and hardware. Given the importance of the Whittle jet to Britain, it made a substantial commitment of key personnel to the transfer of special knowledge to the USA. With the engine and drawings came three engineers to work with the GE team at Lynn: D.N. Walker, one of the Power Jets senior engineers, G.B. Bozzoni, a highly skilled engine fitter and builder there, and Flight Sergeant J.A. King, who had been seconded to Power Jets from the RAF.

Walker returned to Power Jets after two months. However, Bozzoni and King spent five months at Lynn, returning in March 1942. Whittle himself also went out to the USA in June 1942 for almost two and a half months with frequent contacts with the GE team. Coming after Whittle's nervous troubles and breakdown, the trip to the USA seems, on the whole, to have been a welcome break for him, and he wrote back to the UK that 'I get on very well indeed with the engineers over here, because they have both enthusiasm and ability, which is a pleasant change from some we know of.' However, he began again to experience feelings of exhaustion and strain and was induced to relax for a while in California, recalling that, at one Beverly Hills party, he was induced to 'relax' so well that, at 3 a.m., he dived into the pool to join other revellers and ruined his wristwatch. Whittle's contribution to the GE work seems to have been sincerely appreciated, and on his return Air Marshal Hill, the senior officer representing the RAF in America, wrote to Air Marshal Linnell, 'Whittle's visit has been of major significance in promoting closer understanding … . It has made a direct contribution in accelerating diagnosis of the causes of a number of teething troubles experienced by the G.E.C. … . He has won great respect from American engineers … .'

At GE, the development of the Whittle engine was run by Donald F. 'Truly' Warner, a 'top engineer from the turbo-supercharger department', with the first US-built version, the Type I, being equivalent to the Power Jets W.2B. However, GE had re-drawn the designs to suit American machine shop practice, strengthened the impeller blades, and added an automatic control system. The GE engine also used a superior, US-sourced alloy for the turbine blades from the Haynes Stellite Company and first ran on 18 April 1942, approximately six months from receipt of the drawings. At the same time, the Bell Aircraft Corporation was commissioned to produce the Bell XP-59A – a twin-engined aircraft roughly equivalent to the Gloster Meteor. With two GE I-A engines this made the first jet flight at Muroc Lake on 2 October 1942 just under a year from receipt of the Power Jets drawings. This was a truly impressive engineering performance which reflected American expertise, energy and capital. Whittle reckoned that the jet engine test facilities being built at GE were costing about twelve times what had been spent on equivalent facilities for Power Jets. (Incidentally, the original test cell for GE's first Whittle-type engines is still preserved at the Lynn site.)

Sadly, the engineering and personal troubles in the jet programme in Britain, which have been discussed, meant that the W.2B engines were not considered adequate for flight there for some time after this, and in consequence the Gloster F.9/40 (Meteor) prototype first flew with de Havilland Goblin engines in March 1943, five months after the US achievement of its first jet flight, while the W.2B-powered prototype finally flew in June 1943.

GE continued to develop the Whittle concept and came, like Rover and Rolls-Royce, to adopt a straight through combustion chamber layout. The resulting engine, the GE 1-40, first ran in January 1944 and initially developed between 3,800 and 4,500 lb thrust, and stimulated Rolls-Royce to start work on the Nene, intended as a 5,000 lb thrust engine. However, while Rolls-Royce and its licensees received only enough orders to build 4,375 Nenes, the GE 1-40 was passed for manufacture to the Allison company, which produced some 15,500 engines (under the service designation J33), principally to power aircraft like the Lockheed F-80 Shooting Star and the Lockheed F-94 Starfire. J33 production lasted until 1959. The size and length of this production run is eloquent comment on the difference in the economics of development and manufacture in the USA and Britain. (Even the Rolls-Royce Derwent, Rolls-Royce's most successful Whittle-type engine, though built also by Pratt & Whitney in the US, as well as in France, Belgium, Canada and Australia, did not approach the figure for the GE engine, reaching a total production of 9,749 units.)

However, for GE, the I-40/J33 was the end of the line for the development of the Whittle concept. The same air force requirement for a 4,000 lb thrust engine which had brought on the 1-40 also suggested to GE an adaptation of the earlier axial flow turboprop, the TG-100, which had been developed at Schenectady at the request of Arnold's Special Committee on Jet Propulsion. This pure jet engine, designated the TG-180, entered service as the J35, again manufactured by Allison. As the first axial flow engine to be used by the US Air Force it proved to be a significant post-war engine, powering aircraft such as the Republic F-84 Thunderjet, the Northrop F-89 Scorpion and the experimental Boeing XB-47 bomber.

As the war ended GE decided to concentrate gas turbine work at Lynn under the management of

Harold D. Kelsey and to concentrate on axial flow engines. It also set out to win production orders rather than just design and development contracts. GE management predicted that by 1950 the total gas turbine market would be worth $35 million, of which the company might win 25 per cent. In fact, by 1950, GE's share alone was worth $350 million, with their J-47 engine, which powered the North American F-86 Sabre, contributing much to their earnings. Today, like Rolls-Royce, GE is one of the three major global players in the civil and military jet engine market. Both owe their start in the business to the impetus provided by the wartime Whittle jet engine programme.

• Chapter 7 •

The Nationalisation of Power Jets

The take-over of the Rover gas turbine programme by Rolls-Royce transformed the outlook for the series production of the Whittle engine, but the position of Power Jets still remained an irritant within MAP. The outlook of MAP officials was also changing with the arrival of Stafford Cripps as Minister in late 1942. As a committed socialist and a former manager of a munitions plant in the First World War, with a brilliant inter-war legal career, Cripps was important to the Coalition government and a clever choice by Churchill, who captured a sense of Cripps' clearly worn high-minded morality with the quip, 'there but for the Grace of God goes God'. Cripps was deeply impressed by the now highly refined production planning procedures at MAP, alarming the planners themselves, who knew the uncertainties they were dealing with, by commenting to John Jewkes and Ely Devons that 'there is no limit, in theory, is there, to central planning?' The MAP organisation he found for running the aircraft industry appears to have chimed perfectly with his own highly dirigiste approach and from 1943 he began to contribute to government planning for reconstruction, arguing that the huge capital investment in MAP facilities should be used for a major step forward in industrial regeneration and to ensure full employment in the post-war era. 'I conceive', he wrote, 'that a special responsibility rests on MAP to make its capital assets available as the basis of a large engineering industry ... substantially in excess of that existing before the war, and representing the minimum retrogression from that which now exists.'

Cripps also set in train major initiatives for increasing the competitiveness and the technical base of the aircraft industry for the post-war era. Firstly, he consulted with senior aeronautical scientists to launch a new 'National Aeronautical Establishment' conceived on a vast scale and intended to provide wind tunnels and other facilities to serve the industry for the next ten to fifteen years. The cost was to be about £18 million – some £500 million in today's prices. He also energetically backed moves to found a purely aeronautical post-graduate college of aeronautical engineering (today, Cranfield University). All this relates closely to the fate of Power Jets.

On a day-to-day basis, Bulman remained among the most formidable opponents and objected to the company having 'the autonomy of a private firm, whilst being entirely supported on both current and capital accounts by M.A.P.', arguing that, in its relationship with the producer firms, Power Jets was, in effect, functioning like a government research establishment and it should therefore be brought within the Engine Department of the RAE. (He also instigated an enquiry into the salaries and expenses of the Power Jets employees, although this found no irregularities.)

Now, in addition to the dissatisfaction within MAP, there was added internal company criticism of the management which was expressed by workers at Power Jets to Stafford Cripps when he visited the Whetstone factory in 1943. Fielden's account of this was that 'Cripps must have realised straight away what the situation was – the management structure *non est.*' Tinling appeared out of his depth and dried up giving the vote of thanks; the works manager, 'with surprising naivety', called for a response to Cripps' speech and a shop steward took the opportunity to say that the Power Jets management was no good, 'which was true'. As a result an investigation into Power Jets' management was conducted by Eric Mensforth, Chief Production Adviser at MAP, and the MAP Regional Controller which recommended that the management be strengthened.

The question of the ongoing Treasury finance for Power Jets was also on Cripps' mind and he initiated discussions as to how the benefit of this public expenditure could be secured for the nation. His own view was that:

> *Since substantially the whole cost of developing the gas turbine engine had been defrayed by the Government, arrangements ought to be made to secure the benefits of this development as Government property. Power Jets were entitled to be rewarded for having worked on the original idea and for having backed it when the Air Ministry were lukewarm about it, but for this and nothing more.*

The outcome of these discussions was a proposal for the nationalisation of Power Jets. This action was certainly problematic in view of the anxiety about upsetting the fragile jet programme in any way and feelings over the moral and commercial rights of Whittle and his original associates in the venture. Indeed, it seems unlikely that so final a step would have been taken at all, were it not for the constellation of forces pushing in this direction, and, in particular, the fact that the nationalisation was seen as being able to satisfy two quite different policy aims. As we have seen, for some MAP officials the administrative relationship with Power Jets appeared deeply unsatisfactory, and the attitude to Power Jets of the 'hard line' faction within the Ministry can be seen in a note from Sir Lindsay Scott, successor to Linnell as CRD, who expressed the view a little later that:

> *We poured money into that company far in excess of its own capital until it became merely an inadequately organised and controlled organ of public policy and nationalisation, effected at the cost of relatively heavy compensation to private interests, became inevitable.*

Stafford Cripps was concerned about the ethical situation regarding Power Jets and public investment in the engine, but he seems, in reaching the final decision to nationalise the company, to have been also swayed by his strong desire to create new institutions to advance the technological base of the whole British aircraft industry. Thus his thoughts on Power Jets crystallised into the view that an equivalent to the projected new National Aeronautical Establishment was needed for the emerging gas turbine engine industry. Confirmation that his thinking on this was indeed strategic, and not simply an expedient solution to the continuing problem of Power Jets, can also be gathered from the fact that, some time before nationalisation was proposed, he sought advice on the best structure and role for a jet engine research centre from Harry Ricardo, whose own company at Shoreham was a notable centre for engine development work.

The odd position of the Power Jets company offered a vehicle for an analogous establishment in the new jet engine field. It was clearly a creative concern, but was not set up for quantity production. Cripps seems to have held a genuine respect for Whittle and to have believed that a new organisation could be a vehicle for his abilities. Thus the decision to convert Power Jets into a research establishment is entirely consistent with Cripps' other initiatives for fostering new establishments for education and research to support the aviation industry in the post-war era.

Discussions about the valuation of Power Jets' assets ensued between MAP and the company but agreement could not be reached and on 26 November 1943 Cripps wrote to Bonham Carter that there was 'no basis between us for acquiring the assets of the company'. He proposed therefore to consider 'the other alternative' but, significantly, added that 'we urgently need the plant for general experimental purposes'. A subsequent letter from Sam Brown, an undersecretary in MAP and head of the Capital Finance Department, confirms that there was a clear desire in the mind of Cripps to set up a research centre for the new jet industry.

> *The Minister reached the definite opinion that the national interests demanded the setting up of a Government-owned centre of gas turbine technology. The present stringency of building labour and resources generally renders it quite impossible for the Government now to construct a suitable*

> *new establishment. In these circumstances the Minister was forced to the conclusion that he had no option but to exercise his rights, which are not, I think, in dispute, to retake possession of the facilities occupied by the Company at Whetstone, but constructed wholly at the Government's expense, and which are now and always have been the property of the Crown.*

The decision came, understandably, as a brutal blow to the directors of Power Jets, and the terms that were put left no room for argument. Whittle's record was that Cripps stated:

> *If there was no quick agreement on the sum to be paid, [he] would take the alternative course of taking over possession of all the plant operated by the Company and directing all the labour to a new Government Company, leaving Power Jets in possession of its paper assets, such as patents, etc., and nothing else.*

As Schlaifer succinctly put it, 'the government's chief argument in driving so hard a bargain was that without the government financing after July 1, 1939, which amounted … to about £1,300,000, or over sixty times the private investment, the company could not possibly have succeeded in producing a useable product and must necessarily have been liquidated.' There was also the substantial investment in Rover, amounting to some £1.5 million, which had produced almost no visible return in terms of useable engines and although this expenditure was independent of Power Jets finance it seems likely that it would have influenced MAP thinking on compensation. Thus the total government expenditure by the time of nationalisation of bringing the Whittle jet to production was £2.8 million or about £78 million at 2003 values.

Williams and Tinling each received approximately £46,786 for their 'A stock' while Whittle had waived his financial interest. The actual investors of cash or services (the holders of the B stock) received just over three times their initial investment and the total cost to the government of the acquisition was £135,500.

The investors in the company found themselves in the curious position of having their hopes raised by press announcements of the successful development of the jet aircraft while, almost simultaneously, learning from the directors that MAP had taken steps to nationalise the company. Lawyers for a dissident shareholder began an action against the directors of Power Jets for accepting the government terms, putting the powerful argument that investors had waited patiently for nearly eight years for a return on their investments and that the directors had agreed to this sale 'just at a time when the genius of Group Captain Whittle had given a phenomenal and almost unlimited value to the Company's undertaking and assets including its patent rights.'

However, in reality the company's negotiating position was weak and was set out for shareholders by the directors.

The facilities occupied by your company had been provided largely by the government. The proposed utilisation for a Government-owned centre of gas turbine technology would necessitate the Ministry re-taking possession of the facilities and the continuance of the development [of the jet engine] being carried on by the Ministry; the staff being if necessary removed from the Company under appropriate wartime powers. The result ... would be that your company would perforce remain inactive until the cessation of hostilities and restoration of more normal conditions enabled it to take up once more the development and commercial exploitation of the invention ... .

While your Directors would have liked to have been able to get better terms and have tried their best to do so, they are satisfied that this is the best the Minister will offer and, if not accepted, the Minister will re-take possession of all the important facilities now enjoyed by the Company, operate the patents for the service of the Crown as well as he thinks fit and the shareholders may ultimately find that all, or the greater part, of their money has been lost.

Although the directors of Power Jets were obliged to recognise *force majeure* in these dealings with the MAP the nationalisation was understandably a bitter experience. Williams subsequently stood for Parliament and much later said, 'I went into politics because I was so annoyed with Sir Stafford Cripps that I wanted to wipe the floor with him. Unfortunately he died and had himself cremated, so I couldn't even piss on his grave.'

Whittle was placed in an extraordinary position – converted into a national icon at the same time that his company was taken from him. L.L. Whyte, as we have heard, considered it 'Greek tragedy in the modern world'. On 6 January 1944 the jet engine programme was publicly announced with Whittle credited as the inventor – the same day that the directors of Power Jets were informed that the company was to be nationalised.

Whittle himself believed that he 'triggered off the train of events which led to this result' for in April 1943 he had written to Cripps setting out his view that the whole gas turbine industry should be nationalised. Many subsequent writers have assumed that this action was influential in determining the fate of Power Jets. However, official papers show that the central concerns of Cripps and MAP were the ongoing maintenance of a successful jet programme, acting properly with regard to the state's substantial financial investment, and attempting to establish a research centre which could utilise the talents of the Power Jets team. It is most unlikely that Whittle's letter had any influence on the outcome.

One side effect of nationalisation was to solve at a stroke the almost impossible tangle of rights over intellectual and moral ownership of much that had been done, resulting partly from the dissemination of Whittle's ideas through the Gas Turbine Collaboration Committee, and the patent question, discussion of which had been postponed until after the war. Lord Kings Norton (the government scientist who, as Harold Roxbee Cox, ran the wartime gas turbine programme for the MAP) did not consider that these secondary considerations influenced Cripps. Although he personally felt that the state acquisition of Power Jets was 'a pretty scandalous deal' and that the company was very badly treated, in effect by Cripps, he made the reservation that 'I don't think it was wilful – because he was too good a man.'

• Chapter 8 •

A State-Owned Firm: Power Jets (R&D) Ltd

The nationalisation of Power Jets was far from the simple acquisition of the company by the state. It included a novel, indeed a highly original, element – the merger of all the government scientists in the Engine Experimental Department of the Royal Aircraft Establishment at Farnborough who were working on gas turbines into a new state-owned company, together with the Whittle team. This action certainly underlined the seriousness of Cripps' attempt to create a powerful national research centre for the emerging gas turbine industry, for Farnborough was a national institution with a proud history and a record of success in aeronautical affairs, especially in response to the challenge of the Second World War. To change the status of these government scientists who were active in this new and promising field was virtually without precedent, in an age before privatisation, and, in the context of those times, fully as radical as the actual nationalisation of Power Jets. It is another example of the asymmetry in the existing histories of the jet that the creation of Power Jets (R&D) Ltd has been criticised frequently from the perspective of Whittle and other Power Jets personnel but it is far less well known that the merger was also resented by senior RAE staff working on gas turbines at Farnborough including Hayne Constant, as head of the turbine section, and Sir William Hawthorne, both of whom had been engaged in close liaison with Power Jets. Hawthorne, in fact, described it from the RAE perspective as 'a great tragedy'.

Power Jets (R&D) Ltd was formally established as a government-owned company on 28 April 1944 but although direct government control was intended 'to cut a number of Gordian knots in personal relationships' the ending of Power Jets' independence did not fully resolve the problems in Power Jets' relationships with both the aero engine industry and MAP officials.

George Bulman, who was planning the building of the ambitious new National Aeronautical Establishment (NAE), now established near Bedford, cast doubt on the need for an ongoing research and development role for Power Jets, since 'a major objective [of the NAE] is to bring together aerodynamic and powerplant research.' His opposition certainly reflected the poor opinion of Power Jets which he formed while he was in overall charge of engine development, and he questioned 'whether Power Jets was to continue as the chosen instrument for turbine R&D', when logically it should be absorbed into the NAE, which was beginning to be established on airfields near Bedford. (The facility subsequently became a branch of the RAE, called RAE Bedford, in the post-war era.)

Cripps, however, did not accept this advice and a note to Power Jets from an aide shows the sincerity of the intentions he had declared earlier during nationalisation to secure the talent in the company as a national resource:

> *I am directed by the Minister of Aircraft Production to refer to his decision that you should act as the recognised national establishment for furthering, ... the advancement of knowledge on gas turbine engines ... . The Minister feels that this decision involves the installation at Whetstone, and not in the new National Establishment ... [of] the equipment ... which will be required exclusively for the testing of gas turbine engines.*

This action certainly demonstrates that for Cripps, at least, the nationalisation of Power Jets had not been a punitive act and the genuine commitment to build something worthwhile on the bones of the Whittle concern is visible from the resources that were brought to the company. The new board, strengthened in accordance with Sir Eric Mensforth's recommendations, was an extremely powerful assembly, comprising Sir William Stanier (the distinguished steam locomotive designer and chief engineer to the LMS railway, then serving as Scientific Adviser to the Minister of Aircraft Production), Harry Ricardo and, from MAP, Edwin Plowden, now Chief Executive, and Sam Brown, head of the Capital Finance Department. Williams and Tinling continued as directors while Roxbee Cox, who had managed to get the trust of all the factions in the industry and the Ministry in a unique way, became Chairman and Managing Director. The contribution of this really first-class managerial and engineering talent to the new concern has never been discussed and underlines the real national commitment to capitalise on the jet as a major British innovation.

Power Jets R&D Ltd was now a sizeable facility: it occupied, at Whetstone, a factory of 80,000 square feet (8,000 square metres), and expansion to 120,000 square feet was planned. It was well equipped with both machine and capital plant. The powerful and expensive research equipment included a 6,000 hp steam turbine driving a compressor to produce the powerful air supply needed to test components for jet engines, and installation of three wind tunnels for testing 'cascades' of turbine or compressor blades was in train, as well as the manufacture of a supersonic tunnel. The former RAE turbine facilities at Pyestock, near Farnborough, were also kept on and these included a 4,000 hp electrically driven compressor. The company had a flight test section at Bruntingthorpe airfield, near Whetstone, with two Lancasters and a Wellington bomber, adapted as 'flying test beds' to carry jet engines aloft for airborne tests, and two Meteor jet-engined fighters. Over the first year of operation the staffing of the establishment rose from 1,086 to 1,327 employees, while the cost of the first year's experimental work, given in the first annual report in April 1945, was estimated at £600,000.

These figures and resources make it eminently clear that MAP initially acted completely in the spirit of Cripps' plan for Power Jets (R&D) Ltd to become 'the recognised national establishment for gas turbine engines'. Power Jets was not nationalised to smother it to death stealthily. Sadly, though, the role envisaged for the new government-owned company did not properly materialise, due again to problems of personality, communication and understanding.

• Chapter 9 •

The End of Power Jets: Formation of the National Gas Turbine Establishment

In spite of the problems of the jet engine programme and the past arguments between Rover, Power Jets and MAP, the Whittle team was being offered a genuine opportunity to remain at the centre of national gas turbine research, with Whittle still in a central role. Stafford Cripps, Wilfrid Freeman and MAP advisers had considered the issue of Whittle and Power Jets with great seriousness and some magnanimity and it had seemed to them that this new research and consultancy role would be a genuinely useful and fulfilling way of utilising the Whittle team's talents. There would certainly have been no point in supporting it so fully with plant and resources, in wartime, if the creation of the new firm had been considered merely as a sop to Whittle.

Sadly, this scheme did not work out. Having been at the heart of great events – a revolution in aviation, as they saw it themselves at the time – the Power Jets engineers could not adapt to the role of government establishment researchers. One of them suggested that now 'productive development is to be deprecated and neurasthenic intensive research [is] ... the key to heaven. ... An organisation which claims to do development must do research; an organisation which claims to do research (only) is merely a harbour for cranks, inefficient people, people who are afraid of life and misfits.'

Whittle clung tenaciously to his plan to design and build complete engines up to the prototype stage, whereupon, he imagined, the major companies would take up the designs and productionise them – a curious model to cling to after the Rover experience. Whittle, in fact, nurtured a hope that the company 'would retain its dominant position as engine designers' but, by then, this was completely unrealistic, and showed a failure to understand how far the new turbine teams in the mainstream engine firms had come on. Their progress had indisputably relied on the enormous kick-start provided by free dissemination of Power Jets technique through the GTCC and the RAE but Rolls-Royce, for example, considered that they now understood the design of Whittle-type centrifugal compressor engines. They were looking forward to a new generation of slimmer axial flow engines using compressor design research in part emanating from the RAE, and from A.A. Griffith, whom they now employed. As evidence of their competence with Whittle-type centrifugal engines the Rolls-Royce team, largely based around Adrian Lombard and the former Rover team at Barnoldswick, had recently built a new, large jet engine, the Nene, which ran in October 1944, just six months after the project was initiated, rapidly achieving its design target of 4,500 lb thrust with few mechanical problems. It was the highest-powered jet at the time in the world. Ironically, the Nene was conceptually close to the Rolls-Royce RB 37 (Derwent) which was derived from the Rover team's original and contentious 'straight through' B 26 engine – the cause of so much past acrimony between Rover, Power Jets and MAP.

Whittle's model of design and development was strongly resisted by the aero engine industry, particularly by Rolls-Royce and de Havilland. This kind of prototype manufacture was unlike anything done in government establishments and added to the anomalous position of the company. The opposition of the other firms to this was, no doubt, largely based on commercial fears ('We're going to grind Power Jets underground' was one remark reported from the GTCC). They were opposed to competition with a government firm and threatened not to collaborate with Power Jets if this scheme was followed. However, there were also strong elements of personal and institutional pride at stake. The actual designers and managers of these companies were not disposed to place themselves under the direction of an outside design body – particularly Whittle's firm. After all, Rolls-Royce had rescued the engine from the Power Jets–Rover debacle and had, throughout the war, shown that its own engineering judgement was quite strikingly good. The company had successfully resisted the notion of overriding control of the jet programme by Whittle in 1943 and it was not credible that they would now accept it in 1945 when their own understanding of the gas turbine had increased so enormously. De Havilland too, as we have seen, had a notable success with their Goblin, giving some 3,000 lb thrust, in the de Havilland Vampire fighter, and were planning a successor.

G.B.R. Fielden felt that Whittle could have retained a useful role in the industry if he had been willing for Power Jets to be transformed into a 'Ricardo type' consultancy but that 'the idea of Power Jets at the centre of a spider's web in the post-war jet industry was a total illusion.' (As previously noted, Harry Ricardo had established a consultancy at Shoreham shortly after the First World War, carrying out a huge range of research and development projects on internal combustion engines of all types for British and international clients. It still survives as a respected and technologically adept company.) As for the ambition to manufacture engines, it has to be recognised that by the late 1930s Rolls-Royce and Bristol were at the absolute pinnacle of mechanical engineering, and in retrospect the missed opportunity for Power Jets, and for Whittle personally, having himself thought up all the possible mutations of the jet engine, was the failure to team up at the right moment with an established engine builder. Sadly, by the time the knowledge of Power Jets advances and technique had been spread through the wartime industry, at the request of government, their particular bargaining advantage had been dissipated.

The former RAE scientists were also against prototype production. William Hawthorne, from his background as a government scientist, recalled much later that from 1941 'I felt very strongly that as soon as a government establishment goes too far into the new design or invention business it tends to lose its ability to help others and its credibility as an adviser.' That general view, from respected government research scientists such as Constant, Hawthorne and Farren, must certainly have permeated into Ministry of Supply thinking. Early in 1946 moves began to end the existence of Power Jets (R&D) Ltd and to put gas turbine research under the same kind of government establishment structure as that used for research into aerodynamics, radar, guided weapons and so on. The pattern was that these establishments, like RAE Farnborough, or the new guided weapons research establishment at Westcott in Buckinghamshire, developed advanced concepts which they passed to industry, and then analysed and helped refine the projects as the firms developed them. Each establishment had a dedicated site or sites, and was staffed by full-time government scientists, technicians and administrators on carefully defined scales of responsibility and pay.

Unlike these establishments, officials felt that, in spite of nationalisation, Power Jets (R&D) Ltd was still 'an elaborate organisation with an elaborate relationship with the Ministry of Aircraft Production', and the process for bringing programmes into being was characterised as 'somewhat complicated and even mysterious'. Also significant is the comment in the official history that 'Sir Stafford Cripps, the chief architect of the existing arrangement, had left the Ministry.' To the permanent staff there, 'the dilemma of public financial backing without public administrative control' was a continuing concern.

Nevertheless, Kings Norton considered that 'the idea that Power Jets should only do research [was] a damn silly idea. ... The policy towards Power Jets was that PJ should only do pure research – not even to build prototypes – a damn silly idea. The very able staff of PJ resented this and started to leave – utterly disillusioned. I was too, but I had to earn my living.'

Kings Norton believed that 'the immense power of Hives [at Rolls-Royce] forced the government to prevent Power Jets making engines. ... There was a belief among the "brass hats" that Whittle did not understand production [and] no consideration was given to giving him what he wanted, [but] he was brilliant – he could have done anything.' More poignantly, he remarked, 'I was on Whittle's side throughout, but I wasn't strong enough to get him what he wanted', and also, 'Whittle and I have always been friends, but I was the instrument of authority and he disagreed with authority.'

Kings Norton felt that Hives held 'the sincere belief' that building engines was the job of industry, not the job of the government. Whether his views were conclusive is difficult to judge. It seems that events had moved on too far for Power Jets to still try to become established as a producer. If things had gone better from 1940 to 1943 its position as a design authority might have been unassailable.

Clearly there was little general sense that the unique constitution of Power Jets (R&D) Ltd was, by then, providing any special contribution to the aero engine industry, and it was formally converted to the National Gas Turbine Establishment on 1 July 1946. Sixteen Power Jets engineers resigned (Whittle himself had resigned earlier in the year), and although this provoked a question in Parliament, the residue of Power Jets passed from public view without further protest or comment.

• Chapter 10 •

Jet Modernity: The Comet

As reconstruction plans gathered pace from 1943, the place of civil aviation in post-war British industry became a pressing issue for government and the industry. It was felt the huge national investment in new technique, new production and invention must surely be an asset that could be transferred to civil production and to exports.

Lord Brabazon, a former Minister of Aircraft Production and famous pioneer aviator, was asked to form a committee to study the prospects for British civil aircraft (technically, there were two successive Brabazon committees looking at this issue). This episode is well known, in a general way, and has often been represented as an unrealistic or insufficiently serious attack on the problem. For instance, the Bristol aircraft designer, Archibald Russell, later opined that 'Brab had loads of charisma. ... One might reflect that the hero of such experience and reputation, with all at his command, ought to have won greater success', adding that the committee's first action (defining the requirement for the future Bristol Brabazon de luxe transatlantic airliner) was 'a high dive into the deep end without looking to see if there was any water there.' To historian Keith Hayward the Brabazon episode was 'deeply flawed in conception and implementation'.

Unfortunately the problems of the largest and most glamorous aircraft proposed by the committee – the overly ambitious de luxe transatlantic Bristol Brabazon airliner – have coloured views about the committee's work, but its general orientation and many of its recommendations for other aircraft were sound. In fact the Brabazon team really worked with considerable depth and seriousness, trying, for example, to understand the special features of American civil types which made them attractive to operators, for Brabazon himself had woken up to 'the astonishing efficiency of American civil aviation' in 1934 'when the performance of the Douglas in the race to Melbourne opened our eyes'. (It came a close second to the winning de Havilland Comet (a special racer), flying the normal airline route with six passengers and 400 lb of mail.)

The failures of the late-war British civil aircraft programme to achieve all that was hoped for it were almost entirely due to factors outside the Brabazon committee's control.

The Brabazon programme produced a list of suggested aircraft types covering the range of expected route types and passenger numbers and drew up detailed specifications for each. The design criteria, in particular, were highly considered. They specified all the relevant performance and safety criteria such as speed, rate of climb, behaviour with one engine shut down and so on, but went beyond that to define aircraft which would be competitive in the international airline market against American competition. The committee drew attention, for example, to the need to reduce the man-hours required for engine changes and a wide range of service operations. Passenger conditions were also specified, with the air conditioning on long-haul aircraft 'to provide no less than 60 lb of air per minute. ... The ventilation of the lavatory compartment shall be such that any odours which originate in them will not be admitted to any other compartment.' The targets for temperature, pressure and the noise level ('a maximum of 60 phons') were all set out.

It must be noted that this British civil programme took place against the background of a rapidly spreading American aviation network and an American industry that was burgeoning with advanced new transport designs. It was quickly evident that Britain was going to be outpaced and British negotiators, including Lord Beaverbrook (as Chairman of the Commercial Air Transport Committee), tried to slow the pace of negotiations on international air regulation with Adolf Berle, the American State Department's civil aviation representative, suggesting, in August 1944, 'a postponement of your project for moving out onto the civil air routes of the world. Instead, we request an International Air Conference.'

Berle appealed in the spirit of internationalism and free trade, noting that:

> *In many parts of the world it is now obvious that the war area is receding and civil needs are ... reasserting themselves. ... The highest considerations of humanity and common sense as well as the inherent interest of establishing ... normal commercial life dictates the extension of civil air travel.*

Even if a conference were to produce complete agreement, Berle argued, it would take time to implement, and he asked for an immediate interim arrangement to be set up.

However, these difficult discussions were taking place in a pre-existing mood of suspicion between Britain and the USA over the air transport services that were being provided during the war, for such services were seen as laying the foundations for post-war airline operations, through their role both in training aircrew and in allowing the development of the necessary infrastructure such as communications and navigation aids, airport equipment and so on. In particular, the extension of Pan American into areas previously opened up and operated by

Imperial Airways gave particular cause for mistrust in Britain.

However, the diplomatic contest was to prove more or less irrelevant to British aspirations in civil aviation for, while Britain had considerable bargaining power for negotiations over landing rights and overflying across the world, as lead member of the Commonwealth, the prospects for manufacturing competitive British civil aircraft and for organising efficient airline services were bleak. In 1943, American airlines were operating 300,000 route miles compared with 72,000 operated together by RAF Transport Command and British airlines. The Civil Air Transport Committee, chaired by Beaverbrook, listed the advantages of the USA in civil air transport. These included an output of 400 airliners a month, with nine types in production and thirteen multi-engined types under development, whereas Britain had none at all in production. The United States was also researching the problems of civil airline operation through its Civil Aeronautics Board and the Civil Aeronautics Administration – government bodies with a combined staffing of 9,000 people. The committee noted that:

Britain has had no opportunity or manpower available for a detailed study of these questions. ... The Americans are operating more than 1,000 transport aircraft on regular services. We are operating 250, mostly American types. Nineteen separate civil airlines are operating inside and outside the USA and ... have gained a great wealth of experience. We have only one company and hence only one line of experience.

At the same time British industry was unable to make any significant progress on the promised new civil types for, even as the war drew to a close, it was to prove extraordinarily difficult to prise adequate industrial and design capacity away from the RAF.

Thus Beaverbrook wrote to Sir Archibald Sinclair, Secretary of State at the Air Ministry, in August 1944, noting that:

What we need at once is 50 Lancasters converted for long-range transport operations. ... During my chairmanship of the C.A.T. [Civil Air Transport] committee there has been plenty of hope and expectation but no aircraft. If we can provide aircraft now there is still a hope for British Civil Aviation.

The Air Ministry expressed sympathy but suggested, from a security point of view, that Britain could not afford to end the war with 'a second-rate air force'. The MAP countered that as secrecy was not a problem the armed forces could be combed for 'alien draughtsmen' (refugees from friendly countries), although those trying to implement the civil programme objected that good new civil aircraft could only be designed by experienced designers from the industry. An Air Ministry official noted that although about 260 draughtsmen had been found from various sources specifically for work on civil types, only 60 of these were actually engaged on civil work 'because inexperienced men were useless for work in the early stages of design at which the civil types now were.' Nevertheless, the Air Ministry (representing the RAF position) refused to agree to the MAP suggestion that urgent civil types should enjoy equal priority with those military types which were not regarded as essential to front-line combat operations.

The comments of Sir William Hildred, Director General of Civil Aviation and a member of the Brabazon committee, show the frustration of those who wished to see Britain make a start in air transport again. He queried whether all of the huge amount of current military work was useful, remarking that 'I have no wish to pry into military secrets but there must surely be some military work which could now be set on one side ... What ... is the need for the Windsor or the Buckingham?' He observed that:

> *The RAF ... intend to cling to the whole of the present aircraft design staff. ... they are going to*

use a lot of it (and indeed are doing so now) for post-war military work. ... They are not entitled to do this ... they will have to face Parliament, the Treasury and the taxpayer ... clinging like this ... to the detriment of civil work. [We are faced] with an absolute impossibility. Get 500 trained men from somewhere and then you can have your Brabazon designs. The situation ... cannot be met by half measures or shilly-shallying. This is a matter for the War Cabinet.

However, in the mood of the times, the requirements of the RAF proved too pressing and only piecemeal design efforts were allocated at the firms to civil types. In fact the stated requirements for draughtsmen and designers on civil projects would have been quite inadequate even if they had been met in full, and are a measure of the slight unreality of all the initiatives to put British civil production on a level comparable to the American. Against the worthy desire to release 500 trained men to be spread over six Brabazon types can be set the observation of Sir Roy Fedden, the former Bristol engine chief, who found in 1943, at the Lockheed factory in California, 500 staff working on the design of the new Constellation airliner alone. At the peak of the design effort the total for that aircraft had been 700 men with 1.7 million drawing office hours expended on it.

Some consider Fedden a partial witness, as a technocratic zealot and tireless campaigner for his industry, but numerous observers from government, the RAF and elsewhere, brought by war procurement work into contact with the American aircraft industry, commented on the disparity with Britain. There were also reservations both inside and outside MAP about the resolve of the aircraft manufacturers to act energetically. Lord Beaverbrook, for example, while representing British interests during the international civil air transport negotiations, observed at the same time that 'this industry looks as if it's a hotbed of cold feet'.

Out of this period of frustration and confusion the idea of a jet-powered airliner emerged from de Havilland, intended to exploit their wartime jet engine success with the Goblin engine. Initially proposed as a three-engined jet-powered mail plane with negligible passenger accommodation, and derived conceptually from the twin-boom Vampire fighter, it gradually transmuted into a more conventional fully-fledged four-engined airliner proposal to be powered by the de Havilland Ghost, the successor to the Goblin. In the circumstances, this was an imaginative idea from de Havilland which gained powerful support from Lord Brabazon and found resonance with the British Overseas Airways Corporation (BOAC). Against the background of the other stalled civil Brabazon projects the Comet appeared as a beacon of hope and was coming to represent deeper national aspirations. In June 1944 Brabazon made a personal recommendation for it to the Cabinet with the powerful and appealing argument that 'its appearance ahead of any rivals would be a timely reminder of the pioneering work done on jet propulsion in this country.' He also commended the valuable operating experience it would provide and 'its great advertising value'. The jet airliner, he suggested, would get 'the best and quickest value for post-Armistice aviation that can be got from the diversion of a small percentage of design effort now. ... We have no hesitation in recommending that such an aircraft should be built.'

Thus the Comet acquired a powerful symbolic status which related closely to national perceptions of the importance of the Whittle jet. The Comet's rapid progress was a response both to these national ambitions and to the almost impenetrable difficulties placed in the way of almost all the other civil programmes by the Air Ministry and the RAF. The well-known structural failures that lay ahead can be seen, in part, as the result of this intransigence, forcing the company, and the project, into a kind of maverick position determined to make progress at a fearsome pace, and produce a new, world-beating design in a single technological jump. The gamble nearly came off.

By 1947 the Comet programme was appearing as an oddity in the national aircraft ordering system, though it was also part of a wider struggle between the airlines and the Ministry of Supply over the policy that the Ministry should be responsible for the ordering of airliners and progressing development on their behalf.

Thus in November 1947, R.G. Strauss (then Minister of Supply) noted that the Prime Minister's directive on ordering procedure was not being followed by the airline corporations. BOAC had refused, in writing, to discuss contracts with his Ministry or to co-operate in the purchase of the Comet. The corporation was negotiating directly with de Havilland and 'there were no prototypes in the accepted sense of the word.' MoS officials saw the Comet as 'an anomaly – a product of the Corporation's orders.' They queried whether the inspection standards were sufficiently high, and commented on the 'uncooperative attitude of the firm [de Havilland]', their insistence on being given a free hand and, presciently, 'the difficulty in finding any way of checking the design standards being used.'

By 1949, the aircraft had been demonstrated at the Farnborough air show and in October de Havilland test pilots John Cunningham and Peter Bugge completed a 3,000-mile flight to Libya and back in eight and three quarter hours. *The Times* called it 'a fascinating glimpse of air travel in the not far distant future'; the Comet had completed the return trip 'well before any present day airliner leaving London at the same time would have arrived at Castel Benito.' It noted: 'since the Comet flew for the first time on July 27 ... its trials have continued with remarkable smoothness and no difficulties have been encountered.' Sixteen Comets had been ordered 'off the drawing board'. The same phrase had often been used a few years earlier as a flattering description of the drive and initiative to productionise the Whittle W.2B engine.

• Chapter 11 •

Comet Failure

The promise of a step-change to jet airline travel arrived at a time when new post-war piston-engined American types such as the Lockheed Constellation and the Douglas DC-4 were already extending the global reach of civil air transport in a quite remarkable way. The Constellation had already pioneered the pressurised cabin, allowing the aircraft to fly higher, above the weather, thus reducing the awful discomfort of long-distance flights in bad weather. However, the Comet was quite special, as we have seen, virtually halving the journey times on some routes. The remorseless noise and vibration of the high-power piston engine gave way to almost unimagined quiet and smoothness, while at jet cruising height of about 35,000 feet the atmosphere was smoother than the air the most modern piston-engined craft ploughed through some 10,000 feet below. Television reporters covering Comet inaugural flights were fond of showing a pencil balanced on end on the dining tray.

The first fare-paying jet flight took place in May 1952, with Comet G-ALYP flying from London to Johannesburg. But into this extraordinary success story came tragedy. Some twenty months later, this same aircraft was lying at the bottom of the Mediterranean. The troubles began in May 1953, when a Comet leaving Calcutta disappeared. The pilot was said to have flown into a tropical storm, although this now seems unlikely. Then two completely inexplicable crashes of Comets leaving Rome followed in 1954. One lay in water too deep for the recovery of wreckage, but the parts from the other, G-ALYP, lying off the island of Elba, yielded substantial amounts of debris from the sea bottom.

The Comets were grounded and extraordinary efforts were put in place to elucidate the cause. Britain's declining, though still contentious, Imperial role gave rise to thoughts of sabotage. Some even, very privately, mused on the lengths to which the American competition might go. The gravity of the situation was emphasised by the unusual step of asking the RAE to take charge of the investigation, rather than the Air Accidents Investigation Branch of the Ministry of Aviation, which had formal responsibility for such enquiries. The reason was, in part, that the Comet disasters were seen not only as a grave injury to national prestige, but also as directly affecting international perceptions of the quality of British aircraft generally, and, therefore, impacting not only on civil aviation and aircraft sales, but on the credibility of Cold War deterrence and defence and the efficacy of Britain's nuclear-armed jet bombers. Writing in April 1954, Sir John Baker, as Controller of Aircraft in the MoS, registered 'the most urgent need ... to resolve and rectify the technical cause of the accidents ... [in the] national interests of the aircraft industry from both the national and strategic viewpoints.'

RAE Farnborough, by then the biggest research establishment in Europe, uniquely had the resources and expertise to construct a rig to repeatedly pressurise an entire Comet fuselage to simulate the cabin pressurisation cycles of take-off, climb and descent, while hydraulic jacks replicated the flight loads on the wings. After some 3,000 simulated 'flights' a fatigue crack, originating at the corner of a window, pushed out a section of the fuselage. This pattern of failure was confirmed by debris from Comet G-ALYP, then being recovered from the seabed off Elba.

Comet was one of the first generation of pressurised passenger aircraft, designed to maintain comfortable cabin pressure while the aircraft flew high in the unbreathable higher atmosphere for speed, smoothness and economy. At the height the failure had occurred, it would have been explosive, tearing the cabin apart.

This pressure differential between inside and outside imposed a considerable structural load on the fuselage structure. This was recognised by designers, and de Havilland had tested their design, as they thought, by a high static excess pressure. However, the official enquiry, an enormous affair, found, following the research of Dr Walker, Head of the Structures Department at Farnborough, that it was not the absolute pressure value which destroyed the Comets but the cycles of load and relaxation as the aircraft climbed and landed in service, initiating a fatigue crack in the corners of the windows or navigation hatch in the cockpit.

In spite of the exhaustive nature of the enquiry, and the substantially correct finding of metal fatigue, two facts about the failures remained submerged for many years. The first was that the metal fatigue, attributed to raised stress at the squared-off window corners, actually had another cause. These window corners, in fact, had quite generous radiuses and should have been adequate. The fatal flaw arose because the structure had been designed to be bonded – glued, in fact – by another wartime British innovation, the Redux process, which could replace riveting and achieve permanent metal-to-metal joints, if clamped and heated adequately while curing. During production, one of the supervising engineers came to the de Havilland chief designer, R.E. Bishop, with concerns that they could not be sure that the window areas were being properly heated and cured, due to the complex shape. Bishop decided that these areas should therefore be reinforced, as he saw it, by normal aircraft riveting after curing. It was this belt and braces riveting, inserted as an afterthought into a high stress area, that caused the failure. The first cracks emanated from the rivet holes in the corner area – not from material in the corner structure itself.

Redux was, by then, being widely used in UK aircraft assembly and was proving to be an excellent process. The continuous bond reduced stress, compared with a series of rivets, and it may well have been considered unhelpful, both commercially and strategically, to dwell on this one specific failure to use it properly.

Another fact not revealed at the enquiry, or for many years afterwards, was that there had been major concerns within RAE about the Comet structure before the crashes. The operating economics of the first jet airliner were expected to be quite marginal and de Havilland had chosen a particularly thin gauge of aluminium skin, to minimise structure weight and to allow the maximum payload. Their calculations showed it would be adequate but RAE was worried. Ironically, their concerns focussed on the ability of the wing skins to tolerate the continual flexing in flight, and even before the crashes structure experts were asking for special tests. In fact, the wing skins proved not to be a problem. The potential cabin pressure problem did not then occur to anyone. In the industry at large, aero engineers kept their own counsel. It was a national disaster, which reflected on all the companies, but Vickers designers, for example, considered that the skin gauge choice was quite unsound. In their aircraft, the minimum thickness allowable was greater, and determined by a simple criterion – 'tool drop' during servicing. They took the view that dropping a spanner or hand tool inside the aircraft was a typical hazard in servicing and the skin thickness had to be adequate to resist this internal blow without denting. The view in the industry, a Vickers engineer commented years later to the author, was that if any of the other major British aircraft companies had built the Comet, the disasters would not have happened.

The result of the investigation was the grounding of the Comet fleet pending structural modifications. Many aircraft were taken into RAF Transport Command, largely to protect de Havilland from financial failure, under a unique arrangement which the Treasury agreed 'as a special measure'.

Opinions differ about how important the Comet disasters were to British aviation. The simple view is that the subsequent four and a half years, during which de Havilland brought forward the larger and completely new (though aesthetically similar) Comet IV, constituted the crucial loss of time. It meant that the re-engineered Comet IV now arrived on airline routes at approximately the same time as the Boeing 707. Those four and a half years, some argue, could have seen the British commercial airline initiative, spearheaded by the Comet flagship, with its then unique jet propulsion, spread through the world.

This view is probably simplistic. It must be remembered that the Comets (even the Comet IV) were quite tiny aircraft in commercial terms, and could not compete in operating economy with the Boeing 707. In keeping with the early British post-war idea of jet flight as a premium 'Blue Ribbon' service the Comet 1 carried only about 40 passengers and the Comet 4 about 70 (rising to 100 with all-economy seating for the Comet 4C). By contrast, the Boeing 707-120 (the first production version) entered service in 1961 with capacity for 180 passengers. Furthermore, in spite of the Comet's impressive speed advantage over piston-engined aircraft in 1954, the Comet IV had lost out by the time the 707 was launched. In the interim, and using their expertise with a new generation of bombers, Boeing had managed to engineer a more highly swept wing, more demanding in design and calculation terms than the Comet's more conservative wing sweep, and this allowed a higher cruising speed of about 620 mph, while the Comets achieved around 500 mph. Cruising speed translated directly into more attractive flight times, but, as the airlines were learning, the surprising reliability of the new turbine engines also meant that the faster aircraft flew more journeys, carried more passengers and earned more revenue.

Against the Comet as a commercial proposition, there was also the issue of basic British ability to compete in world airline markets. In spite of the allure of the jet, and an admitted British technical lead in jet engine technology, US airline operators just did not believe, at that time, in the ability of the UK industry to support its products with parts and service round the world, and the front-rank non-American airlines, like Holland's KLM, clearly felt the same, for they routinely chose American aircraft and engines.

Another point was that de Havilland was, in world terms, a tiny, though admittedly highly innovative, company. It would have been extremely hard for de Havilland to augment its production sufficiently to capitalise on the success of the Comet. A more truly national aircraft sector in Britain could perhaps have exploited the Comet lead. If the hopes held out during the wartime reconstruction debates for the Ministries to take a leading industrial role had come to pass the Ministry of Supply might have been able to broker partnerships, perhaps by teaming de Havilland with another company adept at production such as Vickers. Sadly, the industry overview envisaged for the MoS in the post-war era did not translate into genuinely strategic structural initiatives of this kind and for years the MoS squandered the enormous UK aviation R&D budget on penny numbers of relatively uninspiring commercial aircraft from different makers which differed little in specification and capability from each other.

The lost opportunity, real or imagined, represented by the Comet, bit deep into the national psyche. When cancelling the lacklustre Vickers V1000 (an unconvincing rival to the emerging Boeing 707) in 1955 Reginald Maudling, reflecting, it seems, opinions from the RAE, mused that Britain might have to surrender to American competition on passenger jets during the 1960s, but looked forward (long before plans for Concorde emerged) to the possibility of re-entering the passenger jet field 'with a true supersonic aeroplane' in 1970 or thereabouts. Despite the Comet disasters the ideal of 'jet modernism' remained to fuel the aspirations to build a world-beating passenger jet aeroplane.

• Chapter 12 •

The Leadership So Narrowly Missed

The Ministry of Supply's responsibility for aviation came to an end in 1959 with the formation of a new Ministry of Aviation with Duncan Sandys as Minister. Sandys was said to be a supporter of the supersonic transport project now being canvassed by RAE and industry (which eventually became Concorde). This enthusiasm for a civil supersonic breakthrough was continued by Peter Thorneycroft, Sandys' successor as Minister of Aviation, who, like all the Ministers responsible for the industry since Cripps (both Conservative and Labour), seemed to be acting fully in the spirit of his late-war agenda for reconstruction. To the Cabinet, Thorneycroft noted in October 1961 that: 'I am deeply concerned as to the future of the aircraft industry, for which the Government have accepted a substantial measure of responsibility.' Among other proposals for investment in the industry Thorneycroft anticipated 'whether we like it or not' the almost certain introduction of supersonic air travel and noted that the USA had started investigating an all-steel Mach 3 aircraft, 'but this is an extremely ambitious project, even for them'.

British design research was focussed on an aluminium alloy Mach 2 design, judged to be a more attainable objective. The heating effect of supersonic flight on the alloy airframe would limit the design to this speed, but since the aircraft industry was far more familiar with working in aluminium alloy it appeared the most attractive route. It also seemed possible that the UK could design and produce an aircraft before the more demanding US design. Thorneycroft announced that 'my technical advisers firmly believe that Britain has an opportunity here of gaining the leadership we so narrowly missed with the Comet.'

These advisers were largely senior aeronautical scientists from RAE Farnborough. Intriguingly, the policy in the Duncan Sandys 1957 Defence White Paper, in which he foresaw the end of manned military aviation in favour of the missile, appears to have pushed RAE aerodynamic research towards the supersonic civil transport problem. To many of the front-rank aeronautical scientists the move to supersonic flight seemed an uncontentious and inevitable step in the series of aeronautical progressions that had brought all-metal piston-engined monoplanes, then jet transports and a progressive decrease in journey times.

The frustrating part was that the economics of supersonics remained stubbornly unattractive. In the mid-1950s, the RAE looked at a possible passenger version of the projected Avro 730 supersonic bomber but gave up when it appeared that the transatlantic payload could be only a mere handful of people. Drag, and therefore fuel consumption, was high at supersonic speed, and therefore, to date, though many military aircraft could fly supersonically, they only did so in relatively short bursts. Making a craft which could offer long-distance supersonic cruise was a different matter. However, the outlook changed when the RAE aerodynamicists began to study the characteristics of the narrow delta, prompted by a 1955 survey paper from one of their number, Johanna Weber.

The studies seemed to show that the supersonic 'wave drag' of the delta was low enough, just, to enable an aircraft to be built with adequate range and a sensible passenger load. As the theoretical solutions for the narrow delta began to emerge, the Farnborough scientists adopted it with tremendous enthusiasm. 'We'd solved the classical aeroplane', one remarked, and they relished the challenge of analysing the new types of flow that the new wing and the supersonic regime would bring. Meanwhile the highly persuasive Morien Morgan, as RAE deputy director, 'did the missionary work to get the companies to take an interest in it' and persuaded appropriate civil servants and all the aircraft company chief designers who were prepared to co-operate to join the Supersonic Transport Aircraft Committee (STAC) which began meeting from November 1956.

The political birth of Concorde, and the convergence of British and French aspirations cannot be considered here, but the technical development was intended, at last, to justify the technological ambitions which had been evoked by the jet programme in the war. At RAE, enormous attention was given to testing. No one liked to refer explicitly to the Comet failures, but the memory was deeply ingrained in the institutional consciousness, from its key role in the crash analysis. The fear of hubris, attempting a new leap in technology, to be rewarded with disaster, was too awful to contemplate. Sir James Hamilton, ex-Farnborough aerodynamicist and head of the Concorde division at the Ministry of Aviation, said: 'I was obsessed with safety. This ... was the most tested aircraft of all time. We had rigs for everything.' One was a full-size fatigue test rig, following the approach first devised for Comet, while the amount of testing and design analysis that went into the aeroplane was unprecedented. The Olympus engines developed for it were also a virtuoso effort. No military engines have to do what Concorde engines could still do at the end of their service life, flying supersonically long distance, on a daily basis. They were developed at Bristol, now under the leadership of Stanley Hooker, who had been put in charge of W.2B jet engine production in 1943, when Rolls-Royce took over the programme from Rover, showing the tremendous continuity in British gas turbine engineering.

• Chapter 13 •

New Jets

In the event, supersonic flight did not turn out to be the logical next step in aviation and at the same time that plans for supersonic aircraft were going forward in Europe and the USA, Boeing was investigating designs for a different kind of aeronautical progress – a very large transport aeroplane that became the 747. At the time both steps forward looked almost equally risky. Passengers might not be willing to pay the premium for supersonic flight. At the same time, many airline people were concerned that passengers to fill the huge new 'jumbos' might also be lacking. They foresaw fearsome competition, a re-equipment 'war' and airline failures. In the event, this did happen, to an extent, and ironically, PanAm, the lead customer and driver of the large Boeing aircraft programme, did eventually fail. But as predicted this generation of aircraft forced down the price of jet travel to an unprecedented level, making global journeys an almost commonplace event for many.

The Boeing initiative also evoked 'wide body' proposals from Lockheed and Douglas. What all these aircraft needed was a new engine with about twice the thrust of the normal civil engines then used in aircraft like the Boeing 707. The answer was to be the high bypass or 'fan engine' – a variant of the jet also anticipated by Frank Whittle, in which most of the air ingested by the engine is a 'cold stream' driven around a hard-working central 'core' engine by a large fan at the front. This solution is far more fuel efficient at subsonic speeds, and is also far quieter.

During the 1950s and 1960s Rolls-Royce persistently tried to enter the American market, with little success. In spite of their advanced technology and keen prices, American airlines had an entrenched preference for homegrown equipment, and Rolls-Royce became frustrated that their presence in the competition forced the American engines to be better than they otherwise would have been, but was not bringing substantial orders. The solution appeared to be an all-out effort to win the contract to build engines for at least one of the emerging American wide body aircraft. David Huddie, as managing director of the Aero Engine Division at Rolls-Royce, moved to the USA for the period of the campaign, where he worked tirelessly to win a contract.

Rolls-Royce also had a proposed solution, a 'three-shaft' engine which traced its multiple independent stage concept back to A.A. Griffith's theoretical ideas at RAE and Rolls-Royce and which promised better fuel efficiency than the two-shaft rivals from Pratt & Whitney and General Electric. Finally Huddie triumphed, with an order from Lockheed to supply engines for their new L-1011 TriStar. At the time, this was the biggest aerospace export order ever achieved in Britain. Huddie said later, 'I thought the price was keen – but I thought we would make money.' However, by 1970, the costs of development engineering were proving to be far higher than expected. Another blow which many at Rolls-Royce felt contributed to the problems was the unexpected early death of the deeply experienced Adrian Lombard. The cost estimates went awry and Rolls-Royce, the greatest name in British engineering, was forced into receivership in 1971.

The great miracle was that, after this catastrophe, the government of Edward Heath, which acquired the company, primarily to safeguard defence production, decided to finance a programme to get the RB 211 back on track. Among other veterans of engine development, Stanley Hooker was brought back from retirement to Rolls-Royce, as Technical Director. Within months of the bankruptcy the RB 211 was giving its design thrust of over 40,000 lb on the testbed and soon went into service on the TriStar.

Perhaps the most intriguing part of this story is that the national finance found to underpin RB 211 development, although found under duress, proved to be a superb investment. No one knew it at the time, but gas turbines were entering a period of relative technological maturity. The RB 211 established an excellent and flexible engine architecture which proved to be capable of continual improvement and upgrading over decades. Today, engines in the Rolls-Royce Trent series can offer over 100,000 lb thrust, still based on that architecture, and the company has moved from the position of underdog in the civil field, to winning and holding about a third of the global airline market for engines against the two principle American makers, General Electric and Pratt & Whitney.

In world terms, this is a major achievement, and a major vindication of Whittle's dream, although the industry went through many vicissitudes to get to this stage. It also proved to take far longer to build a global high-tech jet engine industry than any of the late Second World War planners could have anticipated. At the end of the war optimists hoped that UK aerospace industries could reach parity with those in the USA in about five years. In fact, for turbine engines, it took 50 years.

What kept Britain engaged in gas turbine engineering through those decades when so many new British high technology ventures, for example the computer industry and integrated circuit ('chip') manufacture, melted away? Partly, of course, during the Cold War, there was support for the industry for defence reasons. There was, too, the peerless ability of Rolls-Royce, which could always rival the rest of the world for engineering quality and had been given a massive 'pump prime' through wartime development with the new turbine.

There is perhaps something more. The clue comes from L.L. Whyte's uncannily prescient initial impression on meeting Whittle, his sense that 'it was like ... meeting a saint in a much earlier religious epoch. One surrendered to the enchantment of a single-minded personality born to a great task.'

A saint, through exceptional acts, or exceptional sacrifice, inserts an idea into popular consciousness, and inserts it with such force that it resonates down the generations. Whittle was almost consumed by the emotional effort of fighting for and developing the jet, retiring from the RAF in 1948 on medical grounds. It was Whittle's passion, in all senses, that established the national will to treasure and keep faith with the jet.

ENDNOTES

A Jet Without Whittle? Developments in the USA

Would there have been a jet without Whittle? Can we personalise any invention to the degree that jet history has done? Writing, not of Whittle, but of Diesel, Sir Harry Ricardo suggested that 'we are too fond, I think, of crediting a few particular individuals with a monopoly of inventive genius. The world is well stocked with men of scientific knowledge and wide imagination, and it is with no disrespect to the late Dr Diesel that I suggest that, had he never existed, an equally suitable engine to deal with these heavy oils would ... have been developed, and ... at about the same time.' He also remarked: 'Nothing, I think is more distasteful or invidious, than arguing about who invented what and when, for most intelligent people come to much the same conclusion, at much the same time.'

Against this, we should set Kings Norton's long appreciation, made to the Royal Commission on Awards to Inventors in 1948, in which he noted:

> *It may be said that without Whittle the jet propulsion engine and other applications of the gas turbine would have come just the same. They*

would. But they would have come much later. Whittle's work gave the country a technical lead of at least two years. Properly exploited and maintained, this lead should mean that this country can sell its aircraft gas turbines for years to come.

It seems clear that Whittle's jet initiative, in spite of subsequent development difficulties, did give Britain an early launch into the turbine industry. His personality and passionate advocacy for the engine, and even the difficulties in bringing it to birth, all went to impress on industrialists and policymakers like Cripps that this was a lead in invention that Britain should hold on to.

We might, though, also consider the long-term effect of the brief existence of Power Jets and what would have occurred if Whittle had not obtained official support. In such a case intelligence reports of German jet developments in 1942 would, one assumes, have energised a crash programme in which the advanced work of the RAE on axial flow compressors would have been put together with a major aero engine company, such as Rolls-Royce. A viable engine would certainly have resulted, for a type not unlike this hypothetical motor was, in fact, produced as the Metrovick F2 Beryl during the war, arising out of the partnership between RAE and Metropolitan-Vickers.

However, in the absence of a Power Jets–government relationship it seems unlikely that so high a degree of administrative attention would have been focussed on the jet programme. The 'spoon-feeding' of Power Jets, the drawn out negotiations, and the personality of the inventor himself combined to underpin the idea of the jet engine as something exceptional and extraordinarily hard to engineer. Having engineered it, the British state came to regard it as a great prize, to be nurtured for strategic and economic reasons. As a propagandist for the jet, Whittle certainly succeeded, and without him there would also have been fewer specially designated posts within MAP to nursemaid the gas turbine, probably no National Gas Turbine Establishment, and a far less developed national sense of the jet engine as a triumph of British endeavour and technique. In the absence of this overt commitment there would also have been less likelihood of British companies winning a substantial share of global jet engine business.

Another counterfactual question might also be considered: 'What if there had been no war to provoke the development of the jet engine?' In such a case, it seems likely that the gas turbine would have been deferred by ten or fifteen years and, in the civil field, the new American high-altitude pressurised airliners, such as the Lockheed Constellation, would have had a longer life with their turbosupercharged piston engines before the extension of services round the world prepared the ground for another jump in performance, comfort and speed, with a new turbine powerplant. In this case it seems highly probable that the successful commercial development of such a powerplant (as opposed to its 'invention') would have been in the USA, since, during the 1920s and 1930s, American aviation showed a growing technical lead in the civil field and a growing ability for the industry to capitalise new developments. Indeed, this lead continued, in the civil field, even during the war itself, with design and development still ongoing for new types like the Lockheed Constellation and Douglas DC-4. From the 1930s the American aviation industry was undergoing a step-change in the capability of the whole air transport system. It appears unlikely, in the absence of a major war, that finance would have been found in Britain to support a major new development like the jet.

As discussed in Chapter 6, several gas turbine projects were under way in the USA by the late 1930s, none of which, it appears, derived from knowledge of Whittle's experiments. It was also noted that the turbosupercharger department of GE, under the direction of Sanford Moss, had solved the materials and manufacturing problems of turbine blades capable of standing the heat of the piston engine exhaust gas and, partly for this reason, subsequently made a success of the Whittle engine when this was passed to them in 1941. It seems, therefore, that even without the 'free gift' of a Whittle engine and drawings, interest in the gas turbine would soon have developed at GE. To answer the question posed at the beginning of this section, there is a powerful supposition that if rearmament in Britain had not brought government support to Whittle, the practical aircraft gas turbine would have been realised in the USA a little later as part of the broad sweep of improving aeronautical technologies. It seems highly unlikely, in such a case, that Britain would have emerged with a substantial independent gas turbine industry.

The Jet in Germany

The handling of the jet programme in Germany also makes an interesting comparison with Britain and the Whittle team. There, events initially showed a remarkable parallel to those in Britain. A young German physicist and inventor, Hans von Ohain, began work on a jet engine in 1934. Thus in both Britain and Germany the gas turbine work was launched by a lone inventor, outside the mainstream of the aero engine establishment. Ernst Heinkel, a mercurial aviation industrialist and self-propagandist, took up von Ohain and an experimental jet-powered Heinkel aircraft made a rather marginal demonstration flight in 1939. However, Heinkel was an airframe company and the German air ministry was not greatly interested in an individualistic 'sport' development. Helmut Schelp, the visionary official at the air ministry who did most to stimulate the jet programme, then approached the more established engineering-based companies BMW, Daimler-Benz and Junkers.

BMW and Junkers agreed to take up the gas turbine and were given the latest results of axial flow compressor research emanating from Professor Albert Betz and W. Encke at the AVA aerodynamics research institute at Göttingen (in effect, the German equivalent to RAE Farnborough). Moreover, these two firms were 'very rigidly directed by the *Reichsluftfahrtministerium*'. Von Ohain stayed in the jet programme, but not in a controlling capacity, working on a longer-range, second-generation engine, the Heinkel-Hirth 011 turbojet.

Thus in Germany, the greatest weight of support went to established companies, with the original inventor being partly sidelined. The parallel in Britain would have been if, at the outset, Rolls-Royce and, say, Bristol or de Havilland were given the bulk of the development funds and instructions to incorporate the (highly advanced) RAE ideas on compressor design. In fact, as we have seen, the reverse happened. The analytical work of the RAE was put at the service of Whittle's company, which also got the lion's share of attention and development finance, until a crisis was reached in 1943. The independent RAE work on axial flow compressors got rather low priority, going forward not with an aviation company, but with Metropolitan-Vickers, a steam turbine concern with no aviation experience.

After 1943, as we have seen, MAP felt impelled to bring Rolls-Royce fully into the gas turbine programme, although Tizard's initiative in 1941 to broaden it to other companies was perhaps the first implicit acceptance of the fact that Power Jets simply did not have the size, and perhaps the experience, to fully develop the mechanical and the aerodynamic design of the W.2B.

On the basis of this comparison, therefore, it is clear that Whittle got exceptional backing until the programme ran deeply into trouble. Even then, Whittle's personal conviction, the charm to which his collaborators have attested, and, it must be said, his great talent, secured tremendous indulgence from officials, even though doubts were rife as to whether Power Jets could mature into the type of organisation able to bring the engine into production and serve as the ultimate design authority for it. This forbearance almost certainly delayed development.

Nevertheless, as we have seen, Whittle was without peer in propagandising for the jet. By contrast German turbine engineers appeared much more self-effacing. Whittle largely left the gas turbine industry after the war but, in the post-war years, the lead German engineers continued to work in the area, many of them to a considerable age. Hans von Ohain went to the USA under the auspices of 'Operation Paperclip' (the same US initiative that took Wernher von Braun with his V2 team, and many other scientists). He made little capital out of his pioneering jet work and spent many years at the Aeronautical Research Laboratory at the Wright-Patterson Air Force base near Dayton, Ohio, becoming Chief Scientist there. He was subsequently Chief Scientist at another government establishment, the Air Force Aero Propulsion Laboratory, also at Dayton, retiring in 1979.

Anselm Franz, designer of the Junkers Jumo 004 which powered the Messerschmitt 262 fighter, eventually joined the Lycoming company in the USA, directing the development of many gas turbine engines, particularly for helicopters, retiring as vice-president of Avco-Lycoming.

Max Bentele, who worked with von Ohain on the most technologically advanced German turbojet, the Heinkel-Hirth He 011, similarly went to the USA, after a spell in the UK working on a gas turbine for tanks, and was, for many years, in senior positions at Curtiss-Wright Corporation and at Avco-Lycoming. At Curtiss-Wright he became deeply involved with developing the Wankel engine.

The BMW jet team, which had created the BMW 003, escaped this trend of being taken to America. Instead, the leader, Hermann Oestrich, and a proportion of the other experts were 'spirited away', it is said, from Munich, which was in the US occupation zone. They reappeared in Rickenbach, Switzerland, at the newly founded Atelier Technique Aeronautique Rickenbach (ATAR). The team designed a series of major post-war turbojets for the French SNECMA engine company which all bore the designation ATAR.

Jet Histories

The pioneering work of Frank Whittle on the jet engine and the subsequent nationalisation of his company, Power Jets, has been the subject of much attention in the history of technology. Indeed the history of the jet has become an important case study within the subject and is often represented as an example of a particular kind of revolutionary technological change.

At a more popular level, the troubles Whittle encountered were incorporated into British folk myth. Obviously something as 'revolutionary' as the jet had needed an exceptional individual to drive it through. No doubt civil servants had been incompetent and penny-pinching. The whole episode acquired the comforting cadences of a popular fable with a pantomime cast of heroes and villains, supporting the familiar prejudice that Britain is good at inventing but bad at dealing with its exceptionally creative individuals and also bad at converting their work into industrial developments.

After the war Cynthia Keppel, working on the official war history programme, did, in fact, cover the wartime Whittle programme in good detail, referring (though tactfully) to some of the doubts, problems and conflicts, as they appeared from the government side. In fact, the file of the uninterpreted transcripts of contemporary Ministry reports, memoranda and minutes on the jet engine programme which was assembled for this study survives and shows today how guarded Keppel felt she had to be.

Her 'Gas Turbine Narrative' was not published, but was subsequently drawn on for the section on the Whittle venture in the official history, *The Design and Development of Weapons* (1964), though with further weakening of the critical and analytical passages. While not glossing over the difficulties encountered in development, this official history certainly did not report fully the sense of frustration that existed within the MAP towards the project when it ran into difficulties or the resentment and even hostility that developed towards Power Jets personnel. These authors, although having access to a great range of records and with a considerable team to analyse and condense them, were writing at a time when the difficulties were past and when the jet engine was seen not only as a great British success, but also as a potent symbol of a new British identity in which a claimed technological superiority was becoming a defining national quality. There was also a sense, among those involved in, or informed about, policy matters, that Whittle personally had been treated less than fairly.

The public service ethos of the time also seemed to ensure that no individual government official wrote about the Whittle episode from a personal viewpoint, although a valuable account did finally surface, written by George Bulman, who supervised engine development for the Air Ministry and the Ministry of Aircraft Production in the pre-war and wartime years. *An Account of Partnership – Industry, Government and the Aero Engine: the Memoirs of George Purvis Bulman* was published posthumously some 60 years after these events, in 2001.

Understandably enough, Whittle never deviated from the position he set out in his 1953 book, *Jet*, although he amplified the account in interview to produce an expanded version of this interpretation of events entitled *Whittle – the True Story*, by John Golley, in 1987.

Subsequent accounts have done little to analyse the sources of conflict or to discover whether official disenchantment may have been justified. Indeed, the most perceptive and influential recent scholarly work, *The Origins of the Turbojet Revolution* (1980) by Edward W. Constant II, serves, in effect, to underpin Whittle's own contention that the jet engine was so 'revolutionary' that neither piston aero engine companies nor government officials could judge it fairly or bring themselves to give the support that was merited. Constant employed his concept of 'the turbojet revolution' to extend Thomas Kuhn's 1962 analysis of the successions in scientific theories to explain technological change.

Kuhn, of course, came to prominence by the then shocking argument (to scientists) that scientific developments were, in a sense, sociological, rather than the working out of processes driven by pure reason. He argued that accepted theories imply established belief systems that form the foundation of the 'educational initiation that prepares and licenses the student for professional practice', and so practitioners of what he called 'normal science' (i.e. established science) would tend to suppress or disregard new or anomalous results because 'they are necessarily subversive'. Prior to Kuhn, the scientists' own idealised model proposed an orderly flow from new proposition or theory, to testing by experiment or mathematical modelling and then to evaluation and argument in the light of the evidence gathered. But to Kuhn, every existing body of theory represented a belief system to which adherents had conceptual and even emotional commitment and which was embedded in the very structure of their thought.

Practitioners would cling, perhaps vehemently, to such a system until it began to creak under the accumulated weight of anomalies and new results that old theory could not explain. The resulting tectonic shift could be personally and psychologically unsettling – revolutionary even. It would be, in Kuhn's language, 'a paradigm shift'.

In Edward Constant's treatment, the established, highly successful piston aero engine design and manufacturing culture was disrupted and overthrown by the jet rather as the Ptolemaic view of the universe was overthrown by Copernican cosmology. As noted, it is an important element of Kuhnian thought that this type of process will usually be accompanied by a kind of emotional resistance in the minds of practitioners, followed by a psychological step-change in perception 'akin to a conversion experience', and Constant found plenty of evidence, from Whittle's account, to support this argument. However, his analysis is problematical in that it encourages us to view contemporary 'anti-Whittle' opinions and actions merely as manifestations of a resistant 'pre-jet' mentality, and his sophisticated account largely supports Whittle's own view of the difficulties he met.

By contrast, as we have seen, officials and engineers were far more open-minded towards the gas turbine than Constant, or indeed Whittle, acknowledged. In fact, official papers, and letters between the protagonists do not reveal this 'emotional commitment' to the piston engine that a Kuhnian interpretation would imply. For example, in discussion of future high-power engines, Harry Ricardo wrote to Sir Henry Tizard in 1940: 'In the long run the turbine will be more reliable than the reciprocating units.'

NOTES

1. In the post-war period Ralph Dudley Williams changed his name to Rolf Dudley-Williams and embarked on a career in politics. He was elected Member of Parliament for Exeter (1951–6), Parliamentary Private Secretary to the Secretary of State for War (1958), and to the Minister of Agriculture (1960–4). He was created Baronet in 1964. He is referred to by both names depending on context.
2. In the inter-war period Harold Roxbee Cox was a government scientist who worked as a design engineer on the government R101 airship and then did theoretical studies on aircraft structures at Farnborough. After the Second World War he had a distinguished industrial career and was Chairman of the Metal Box company from 1961–7. He was made Lord Kings Norton in 1965. Both names have been used in this book, depending on context.
3. In spite of the disputes with Rover and MAP, life for the development engineers in Power Jets was utterly absorbing and fulfilling. For example, Professor A.G. Smith, formerly Head of the Department of Mechanical Engineering at Nottingham University, thought Power Jets

'a wide awake place ... a wonderful place' and getting his job there in 1942 'the luckiest thing that ever happened to me'. Whittle he found to be 'an accomplished and practical mechanical engineer'.

4. The US Army Air Corps changed its name to the US Army Air Force during the Second World War, and in 1947 became the US Air Force. For simplicity, the service has been referred to as the USAAC in the early period, and subsequently as the US Air Force.

BIBLIOGRAPHY

Records

Aeronautical Research Committee, papers, ledger pages, publications (DERA Library Farnborough; Royal Aeronautical Society Library, London).

AiR, AViA and CAB classes at the Public Record Office, London.

Bailey, W., *The Early Development of the Aircraft Jet Engine*, 1995, unpublished account from a wartime gas turbine worker at RAE, deposited with Royal Aeronautical Society, 1996.

Bonham Carter, Sir Maurice, papers and correspondence relating to Power Jets Limited, Science Museum archives.

Brabazon, Lord (J.T.C. Moore-Brabazon), papers Royal Air Force Museum, Hendon.

Bulman, G.P., unpublished autobiography, ms in the library of the Royal Aeronautical Society, London. Now published in an edited form as *An Account of Partnership – Industry, Government and the Aero Engine*: (cf. p.167) *the Memoirs of George Purvis Bulman*, edited and with a commentary by M.C. Neale (Rolls-Royce Heritage Trust, Derby, 2002).

Dowding, Sir Hugh, papers Royal Air Force Museum, Hendon.

Farren, W.S., and W.G.A. Perring, 'The Functions and Future of the Engine Dept. R.A.E., Aeronautical Research Committee, Engine Sub-Committee', I.C.E. 1532, 7 June 1943.

The Fedden Mission to America, Final Report, Ministry of Aircraft Production, 1943 (Science Museum archives).

The Fedden Mission to Germany (Science Museum archives).

The Future of the Aircraft Industry: Aeronautical Research and Development, First Report by the Aircraft Industry Working Party, 14 April 1958 (under the chairmanship of Sir Thomas Padmore).

The Gas Turbine Collaboration Committee, Minutes and Reports, 1941–50 (Science Museum archives and at the Public Record Office).

Hooker, S.G., *The Future of Air-Breathing Engines in Aviation*, Institution of Mechanical Engineers James Clayton Lecture, 16 December 1959.

Moult, E.S., 'The Development of the Goblin Engine', *Journal of the Royal Aeronautical Society*, Vol. 51, 1947, pp. 655–85.

Report of the Committee of Inquiry into the Aircraft Industry (HMSO, London, 1965), Cmnd 2853 ('The Plowden Report').

Roxbee Cox, H., 'British Aircraft Gas Turbines', *Journal of the Aeronautical Sciences*, February 1946, Vol. 13, No. 2, pp. 53–85.

Royal Aircraft Establishment (RAE) Technical Notes,

Reports and Memoranda (R&Ms), DERA Library, Farnborough.
Tizard, Sir Henry, papers Imperial War Museum.
Whittle, F., 'The Early History of the Whittle Jet Propulsion Gas Turbine', (First James Clayton Lecture), *Proceedings of the Institute of Mechanical Engineers*, 1945, Vol. 152, No. 4, pp. 419–35.

Books

Banks, Air Commodore F.R. (Rod), *I Kept no Diary* (Airlife, Shrewsbury, 1978). (Banks was in charge of engine development at the MAP during the latter part of the war and post-war under the Ministry of Supply.)
Bentele, Max, *Engine Revolutions* (Society of Automotive Engineers, PA, 1991).
Brooks, David S., *Vikings at Waterloo: the wartime work on the Whittle jet engine by the Rover Company* (Rolls-Royce Heritage Trust, Derby, 1997).
Bryant, Chris, *Stafford Cripps: the First Modern Chancellor* (Hodder & Stoughton, London, 1997)
Clarke, Peter, *The Cripps Version: the Life of Sir Stafford Cripps, 1889–1952* (London, 2002).
Connor, Margaret, *Hans von Ohain: Elegance in Flight* (AIAA, Reston VA, 2001).
Cooke, Colin, *The Life of Richard Stafford Cripps* (London, 1957).
Constant II, Edward W., *The Origins of the Turbojet Revolution* (Baltimore, 1980).

Edgerton, David, *England and the Aeroplane* (Manchester, 1991).

Estorick, Eric, *Sir Richard Stafford Cripps* (William Heinemann, London, 1949).

Golley, John, *Whittle: the True Story* (Airlife, Shrewsbury, 1987).

Grierson, John, *Jet Flight* (London, 1945).

Gunston, Bill, *By Jupiter: the Life of Sir Roy Fedden* (R.Ae.Soc., London, 1978).

Hayward, Keith, *The British Aircraft Industry* (Manchester, 1989).

Hancock, W.K., and M.M. Gowing, *British War Economy* (HMSO, London, 1949).

Hennessy, Peter, *Never Again: Britain 1945–1951* (London, 1992).

Holland, Robert, *The Pursuit of Greatness: Britain and the World Role, 1900–1970* (London, 1991).

Hooker, Sir Stanley, *Not much of an Engineer* (Airlife, Shrewsbury, 1984).

Keppel, Cynthia, *Official History: Gas Turbine Narrative*, PRO CAB 102/393.

Lakatos, I., and A. Musgrave (eds), *Criticism and the Growth of Knowledge* (Cambridge, 1970).

Lloyd, Ian, *Rolls-Royce: the Merlin at War* (Macmillan, London, 1978).

Montgomery Hyde, H., *British Air Policy between the Wars, 1918–1939* (Heinemann, London, 1976).

Postan, M.M., D. Hay and J.D. Scott, *Design and Development of Weapons* (HMSO, London, 1964).

Ricardo, Sir Harry, *Memories and Machines* (London, 1968).

Ritchie, Sebastian, *Industry and Air Power: the Expansion of British Aircraft Production, 1935–1941* (Frank Cass, London, 1997).

Robertson, A.J., 'Lord Beaverbrook and the Supply of Aircraft, 1940–1941', in *Business, Banking and Urban History*, ed. Anthony Slaven and Derek H. Aldcroft (John Donald, Edinburgh, 1982).

Russell, Sir Archibald, *A Span of Wings* (Airlife, Shrewsbury, 1992).

Schlaifer, Robert, and S.D. Heron, *The Development of Aircraft Engines and Fuels* (Harvard School of Business Administration, 1950).

Swinton, Lord, *I Remember* (Hutchinson, London, 1948).

Tedder, Lord, *With Prejudice* (Cassell, London, 1966).

Tiratsoo, Nick, *The Attlee Years* (Pinter, London, 1991).

Wegener, Peter P., *The Peenemünde Wind Tunnels, a Memoir* (Yale, 1996).

Whittle, Sir Frank, *Jet* (Frederick Muller, London, 1953).

Whyte, Lancelot Law, *Focus and Diversions* (Braziller, New York, 1963).

Science and Islam
9781785782022

Atom
9781785782053

An Entertainment for Angels
9781785782077
(not available in North America)

Sex, Botany and Empire
9781785782275
(not available in North America)

Knowledge is Power
9781785782367

Turing and the Universal Machine
9781785782381

Frank Whittle and the Invention of the Jet
9781785782411

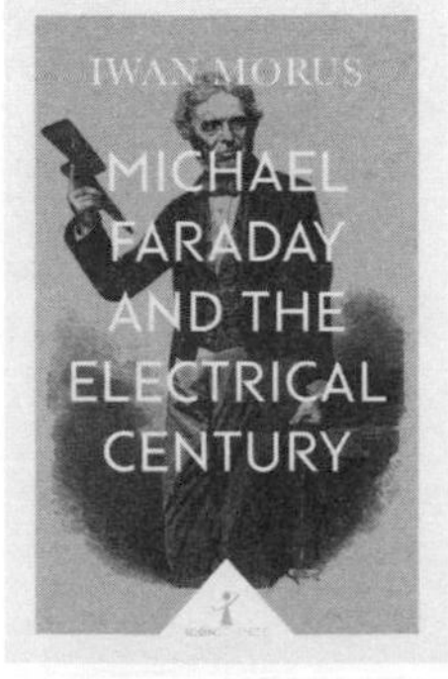

Michael Faraday and the Electrical Century
9781785782671

Moving Heaven and Earth
9781785782695